FISH BIOLOGY

FISH BIOLOGY

PROFESSOR (DR.) ARVIND KUMAR

Environmental Science Research Unit,

Post Graduate Department of Zoology,

S.K.M. University, Dumka - 814 101, Jharkhand, India.

A.P.H. PUBLISHING CORPORATION

ANSARI ROAD, DARYA GANJ

NEW DELHI-110 002

Published by
S.B. Nangia
A P H Publishing Corporation
4435-36/7, Ansari Road, Darya Ganj
New Delhi-110 002
Ph. : 23274050
Email : aphbooks@vsnl.net

2019

Typesetting at
Paragon Computers
New Delhi-110 059

Printed at
Balaji Offset
Navin Shahdara, Delhi-32
Ph.: 22324437

DEDICATED TO MY FATHER

LATE SHRI BRAHMADEO SINHA

A NOTED FREEDOM FIGHTER

A GANDHIAN

A HUMANITARIAN

AN EDUCATIONIST

& AN ENVIRONMENTALIST

PREFACE

Though during the preceding years, India has made substantial progress in fish research and fish production, but it is not satisfactory in comparison to the western countries. So, the task involves intensive research in the field of Ichthyology and Fisheries Science for selection of ideal fish species for aquaculture and their genetic upgrading through selection, cross-breeding and lineage, resolving of ideal feeding conditions and optimal feeds, determination of ideal environment for rapid growth and weight gain as well as effective control of diseases. Equally important is [illegible] research for developing suitable management practices for lentic ecosystem in order to maintain equilibrium in the aquatic ecosystem for optimum Pisciculture.

Life has had an inevitable role in shaping of the history and evolution of this watery planet. During the last 50 years, the earth has experienced and witnessed from many natural disasters. Due to population explosion, large scale [illegible] industries and unscientific disposal of waste [illegible] been threatened. [illegible] world. Various unplanned activities have not only [illegible] have also disrupted the delicate fabric of [illegible] changed the economic and socio-cultural life [illegible] biologist has to learn how to develop [illegible] and promote the fish research for the days to come. [illegible] some of the lesser-known facts, I have ventured to bring out a [illegible] and comprehensive review volume with a hope that the fish biologists [illegible] investigation on their underexplored facts, which [illegible]

My special thanks and appreciation to [illegible] enriched this volume. I wish to express my sincere [illegible] Chancellor, S.K.M. University, [illegible] especially thankful to Professor (Dr.) B.N. Jha [illegible] Dumka for his encouragement. I owe my special [illegible] Chancellor of Sambalpur University, Professor [illegible] Allahabad University, Professor [illegible] Professor N.C. Datta of Calcutta University, Dr. S. [illegible] Mumbai University, Professor S.K. Konar of Kal[illegible] Bhopal University, Professor U.C. Goswami of Gau[illegible] Sambalpur University, Professor R. Natarajan of [illegible] Bangalore University, Professor Ajit Varma of [illegible] Saurashtra University, Professor A.L. Bhatia of [illegible] BHU, Professor S.P. Hosmani of Mysore Uni[illegible] University, Professor K.B. Reddy of Nagarjuna U[illegible] Pondicherry University, Professor B.K. Tiwari of [illegible] University, Professor R. Ramalingam of Annamalai [illegible] Nagaland University, Professor G.K. Kulkarni [illegible] Mehram of Nagpur University, Professor S.K. Ba[illegible] of Manipur University, Professor B.D. Joshi of H[illegible] Faizabad University, Professor K.C. Sharma of A[illegible]

PREFACE

Though during the preceding years, India has made substantial progress in fish research and fish production, but it is not satisfactory in comparison to the western countries. So, the task involves intensive research in the field of Ichthyology and Fisheries Science for selection of ideal fish species for aquaculture and their genetic upgrading through selection, cross-breeding and lineage, resolving of ideal feeding conditions and optimal feeds, determination of ideal environment for rapid growth and weight gain as well as effective control of diseases. Equally important is suitably planned research for developing suitable management practices for lentic ecosystem in order to maintain equilibrium in the aquatic ecosystem for optimum Pisciculture.

Life has had an inevitable role in shaping of environmental evolution of this watery planet. During the last 50 years, the earth has experienced and recovered from many natural disasters. Due to population explosion, large scale of urbanization, mushrooms growth of industries and unscientific disposal of waste materials, the ecobalance of the biosphere has been threatened. Eco-degradation has become a growing challenge to mankind all over the world. Various unplanned activities have not only altered the scenario of the environment but have also disrupted the delicate fabric of man-nature relation. Not only this, they have changed the economic and socio-cultural life of the people. Therefore, each and every fish biologist has to learn how to develop innovative research in order to preserve, conserve and promote the fish research for the days to come. Thus, keeping in view of highlighting some of the lesser-known facts, I have ventured to bring out such an innovative and comprehensive review volume with a hope that the fish biologists may undertake further investigation on their undeciphered facts, which are neglected hitherto.

My special thanks and appreciation go to the scientists whose contributions have enriched this volume. I wish to express my sincere gratitude to Dr. P.C. Hembram Hon'ble Vice Chancellor, S.K.M. University, Dumka who has been a source of constant inspiration. I am especially thankful to Professor (Dr.) B.N. Jha, Hon'ble Pro Vice-Chancellor, S.K. University, Dumka for his encouragement. I owe my special thanks to Professor M.C. Dash, Hon'ble Vice Chancellor of Sambalpur University, Profesor H.R. Singh, Hon'ble Vice Chancellor of Allahabad University, Professor K.C. Pandey, Former Vice Chancellor of Meerut University, Professor N.C. Datta of Calcutta University, Dr. S.N. Kaul of NEERI, Professor U.S. Bagde of Mumbai University, Professor S.K. Konar of Kalyani University, Professor D.K. Balsare of Bhopal University, Professor U.C. Goswami of Gauhati University, Professor P.C. Mishra of Sambalpur University, Professor P. Natarajan of Kerala University, Professor P.S. Murthy of Bangalore University, Professor Ajit Varma of JNU, New Delhi, Professor P.P. Sood of Saurashtra University, Professor A.L. Bhatia of Jaipur University, Professor A.K. Mittal of BHU, Professor S.P. Hosmani of Mysore University, Professor K. Kapoor of Udaipur University, Professor K.B. Reddy of Nagarjuna University, Professor M. Vikram Reddy of Pondicherry University, Professor B.K. Tiwari of NEHU, Professor G. Tripathi of Jodhpur University, Professor R. Ramalingam of Annamalai University, Professor Sharif U. Ahmad of Nagaland University, Professor G.K. Kulkarni of Aurangabad University, Professor S.U. Mehram of Nagpur University, Professor S.K. Battish of PAU, Professor B. Manihar Sharma of Manipur University, Professor B.D. Johsi of Hardwar University, Professor G.C. Pandey of Faizabad University, Professor K.C. Sharma of Ajmer University, Professor A.K. Dobryal of

Garhwal University, Professor Kaushal Kumar of Gorakhpur University, Professor M. Raziuddin of Hazaribag University, Professor Gurdeep Singh of I.S.M., Dhanbad, Dr. P.K. Goel of Karad, Dr. Ram Kumar of Delhi University, Professor S.P. Roy and Shri Tribhuwan Poddar of Bhagalpur University, Bhagalpur.

I also express my deep sense of gratitude of my parents whose blessings have always prompted me to pursue academic activities deeply. I am also thankful to my sweet wife, Kumari Bimla and my two lovely sons, Kumar Pallav Shrivshankaran and Kumar Prasun Ramakrishnan whose natural smiles extended to me relief all through this tiresome endeavour.

Last but not the least, I am also thankful to Mr. S. B. Nangia, Proprietor, APH Publishing Corporation, New Delhi for taking keep interest in bringing out of this book. Finally, I will always remain a debtor to all my well-wishers for their blessings, without which this book would not have come into existence.

Dumka **Professor (Dr.) Arvind Kumar**

CONTENTS

CHAPTER 1

Strategies for Utilization of Biogas Slurry in Fish Pond Manuring

Arvind Kumar

Environmental Science Research Unit, Post Graduate Department of Zoology, S. K. M. University, Dumka - 814 101, Jharkhand.

ABSTRACT

Effects of animal manures (Poultry droppings, biogas slurry and cow dung) on water quality parameters plankton production and survival and growth of fish were studied at PAU. Ludhiana. The water quality parameters were remained in optimum range. Plankton production (Phytoplankton and zooplankton) was observed maximum in ponds manured with biogas slurry. However the growth of *Cyprinus carpio* in terms of percentage increase in length and weight gain was observed maximum in the pond manure with poultry droppings, followed by biogas slurry and cow dung.

Key words: Organic manures, Water quality; Fish production and Fish growth.

INTRODUCTION

India has made impressive and rapid strides in the development of fisheries, particularly in the area of aquaculture. Increasing aquaculture production is no longer the main target but its sustainability is more important. The sustainable aquaculture depends upon the provision of eco-friendly, economically and socially viable culture system.

The supplementary diet contributes to 40-60% of the total production cost in intensive fish culture system. Increasing cost of feed stuffs (grains, fish meal, oil, cakes etc.) used in the preparation of supplementary diet has stressed the scientists to look for some cheaper substitutes. Many wastes and by-product of agriculture, animal husbandry and agro-industries have food value and can be used as pond manures and fish feed ingredients.

Moreover fish is an excellent converter of poor quality ingredients into high level animal protein i.e. the nutrients for natural food of fish (Plankton) which had a significant and positive influence on fish growth. Some fish species also feed directly upon the undigested fractions of these wastes (Schroedar, 1980) which may be low in nutritional value, but microorganisms adhere to them are of high protein value. Further these wastes can also be incorporated into the production of cheap low protein pelleted feed (Devaraj and Keshavappa. 1985).

REVIEW OF LITERATURE

Many organic manures like cow dung, poultry manure, duck manure, pig dug and sewage etc. have been tried. Much work has been done on the productivity potential of cow dung, which is now universally used as the organic manure (Banerjee *et al.* 1969) in pisiculture practice.

Schoonbee *et. al.*, (1979) observed marked fluctuations in dissolved oxygen level in manured ponds (cattle manure) but anerobic conditions did not prevail at any time. Burn and Stickney (1980) reported increase in ammonia with increase in level of poultry waste in the ponds. Sandhu (1982) found that phosphate contents of water increased by manuring (farm yard manure and poultry droppings of 5-17 t/ha/yr).

Yadav and Garg (1992) observed that addition of poultry droppings @ 30-428 kg dry matter/ha and cow dung @ 24 t/ha/yr respectively did not affect the water quality and hydrobiological factors remained within the optimum level.

The animal manure increase the biological productivity of the pond through various pathways, which is judged through the qualitative and quantitative estimation of plankton production (Schroeder. 1980).

The type of manure also influence the plankton production. Biogas slurry has been found to be one of the potential manurial input in enhancing the growth of plankton than cow dung (Rai, 1981, Kalayani and Shetty, 1987, Sehgal *et. al.*, 1991).

Animal manures have direct or indirect effect on the survival and growth of fish. Above 90% survival of *C-carpio* was reported in ponds fertilized with poultry manure (Kaur and Kaur, 1987) whereas 100% survival of *C-carpio* was reported from ponds receiving cowdung and biogas slurry (Kaur *et. al.*, 1987). Dhawan and Toor (1989) recorded maximum growth of *C-mrigala* in ponds treated with poultry droppings followed by cow dung (both @ 20 g/ha/yr) than unmanured pond. Growth rate of 71.15, 48.33, 38.15 and 29.67g of *C-carpio* was reported from ponds receiving poultry droppings (18 t/ha/yr), cow dung (10 t/ha/yr), biogas slurry (19 t/ha/hr) and control (without manure) respectively (Kaur *et al* 1989).

MATERIALS AND METHODS

Experiments were conducted from 19th April 2002 to 25th June, 2002 at the Fisheries Research Complex of Punjab Agricultural University. Ludhiana to study the effect of biogas slurry on the growth of *Cyprinus carpio* (common carp). Physico-chemical parameters of water, plankton production were also analysed to assess whether biogas slurry had any effect on them.

Six cemented tanks (1-6) each with an area of 0.002 hectare (5.85 x 2.77m) and a depth of 1.22 m were selected and manuring was done with cow dung, poultry waste and biogas slurry. Cow dung, poultry waste and biogas slurry were added at desired levels i.e. @ 18 t/ha/yr in a tank (700 g/ha/week). The tanks were stocked with fingerlings of *Cyprinus carpio* (common carp). The different water quality parameters like water temperature, secchi disc transparency, hydrogen ion concentration (pH). dissolved oxygen (DO), alkalinity and free carbon dioxide were analysed. Pond productivity was assessed in terms of plankton production, which includes both phytoplankton and zooplankton, by taking. Plankton samples simultaneously at the time of sampling water for its physico-chemical analysis.

RESULTS AND DISCUSSION

The effects of different manures on water quality parameters, pond productivity and fish growth are shown in Tables 1.1, 1.2 and 1.3 respectively.

Table 1.1 Effect of differetn manures on water quality parameters

Parameters	Days	Manure			Optimum range of water quality parameters
		Biogas Slurry	Cow Dung	Poultry Droppings	
Water temperature (°C)	0	28.75	28.75	27.75	
	20	31.50	31.00	31.50	15-37.5
	40	27.50	27.50	26.50	
	60	33.00	32.00	31.50	
Secchi Disc Transparency (cm)	0	23.25	19.87	24.62	
	20	20.95	18.60	21.75	
	40	17.87	17.05	19.60	
	60	17.45	16.60	19.10	
Hydrogen Ion Concentration (pH)	0	8.87	8.42	8.19	
	20	8.95	8.91	8.63	6.5-9.6
	40	9.07	8.22	8.17	
	60	9.41	8.18	8.59	
Dissolved Oxygen (mg/l)	0	7.40	6.60	5.50	
	20	15.30	12.00	7.90	7.8-9.0
	40	12.40	5.30	5.60	
	60	13.40	10.40	12.90	
Alkalinity (mg/1)	0	275	247	365	
	20	249	231	223	137-230
	40	130	152	158	
	60	166	162	206	
Free Carbon dioxide (mg/l)	0	0	0	0	
	20	0	0	0	15.00
	40	0	8	16	
	60	0	8	0	

Table 1.2 Effect of different manures on pond productivity

Pond productivity	Days	Manures		
		Biogas Slurry	Cow Dung	Poultry Droppings
Phytoplankton (ml/1)	0	34	5	7
	20	10	7	18
	40	27	20	28
	60	22	5	13
Mean		23.25	9.25	16.5
Zooplankton (ml/1)	0	30	10	15
	20	44	22	18
	40	48.52	47	67
	60	35	42	50
Mean		39.38	30.25	37.5

Table 1.3 Effect of different manures on fish growth

Days	Fish growth	Manures		
		Biogas Slurry	Cow Dung	Poultry Droppings
Initial (0 day)	Length (cm)	12.50	14.08	12.30
Final(60 days)		13.76	14.58	14.60
Length gain (%)		10.08	3.55	18.70
Initial (0 day)	Weight (gm)	35	39	38
Final (60 days)		42	42	50
Weight gain (%)		20	7.69	31.57

In the present study, phytoplankton was observed maximum in pond manured with biogas slurry (23.25 ml/1). Similarly zooplankton was observed to be maximum in the pond manured with biogas slurry (39.38 ml/1) followed by poultry droppings (37.5 ml/1) and cow dung (30.25 ml/1). Biogas slurry had been observed to be one of the potential manurial input enhancing the growth of plankton (Phytoplankton and Zooplankton) than cow dung (Rai. 1981. Kalayani and Shetty. 1987). This high value of plankton in pond manured with biogas slurry was because of high nutrients contents of organic manures, which enhance the plankton production.

Fish growth (in terms of percentage increase in length and weight gain) was observed to be maximum in the pond manured with poultry droppings (length gain = 18.7%, weight gain = 31.57%) followed by biogas slurry (length gain= 10.08%), weight gain=20%) and cow dung (length gain=3.55%, weight gain=17.6%). Maximum growth of fish was observed in case of poultry droppings, this has also been observed by Dhawan and Toor (1989).

CONCLUSIONS

Water quality parameters viz. water temperature, pH, dissolved oxygen, alkalinity, secchi disc transparency and free carbon oxide were analysed, which remained in optimum range. The most critical parameter i.e. dissolved oxygen in all the treatments was in optimum range and maximum dissolved oxygen values were in case of biogas slurry. Plankton production (phytoplankton and zooplankton) was observed maximum in ponds manured with biogas slurry followed by poultry droppings and cow dung. The growth of *Cyprinus carpio* in terms of percentage increase in length and weight gain was observed maximum in case of pond manured with poultry droppings (length gain=18.7%, weight gain =31.57%) followed by biogas slurry (length gain=10.08%. weight gain = 20%) and cow dung (length gain = 3.55%. weigh gain = 17.6%).

The organic wastes are available in large quantities all over the world can be used for supplementary diet for fish production. Moreover recycling of organic wastes for fish culture served the dual purpose of cleaning the environment and reaping economic benefits.

REFERENCES

Banerjee, R.K.. G.S. Singit and P. Ray (1969). Some observations on the use of poultry manure as fertilizer in rearing major carp fry. Indian J. Fish. 16 : 29-34.

Burn, R.P. and Stickney, R.P. (1980). Growth of *Tilapia auria* in ponds receiving poultry wastes. Aquaculture 20 : 117-121.

Devaraj, K.V. and Keshavappa. G.Y. (1985). Preliminary studies on poultry-cum-fish culture. *Mysore J Agric. Sci,* 19 (3) : 189-193.

Dhawan, A. and Toor. H.S. (1989). Impact of organic manure and supplementary diet on plankton production and fish growth and feeding of an Indian major carp. *Cirrhina mrigala* (Ham.) in fish ponds. Biol. Waste 29 : 289-298.

Kalayani, R. and Shetty. H.P.C. (1987). Use of bio-gas slurry as a fertilizer for production of carps. *The First Indian Fisheries Forum,* Dec. 4-8.

Kaur, K. and Kaur. S. (1987). Efficacy of poultry droppings in common carp *Cyprianus carpio* (Linn) culture. Effect on survival and growth. *Zoologica Orientalist* (1 & 2): 13-18.

Kaur, K., Kaur. K. and Dhawan, A. (1989). Effects on organic wastes on the growth and ovarian maturation of *Cyprinus carpio. Proc. Nat. Sem. Freshwat Aqua.* Pp. 81-83.

Kaur, K., Sehgal. G.K. and Sehgal. H.S. (1987) Efficiency of biogas slurry in carp. *Cyprinus carpio* var. *communis* (Linn.) culture. Effects on survival and growth. *Wastes* 22 (2) : 139-146.

Rai, S.P. (1981) Role of biogas plant in integrated aquaculture for rural development. In : *Summer Institute on Farming System Integrating Agriculture Live-stock and Fish Culture.* 6th July to 4th Aug. CIFRI. Barrackpore, West Bengal.

Sandhu, .J.S. (1982). Role of organic wastes in composite fish culture. Ph.D. Thesis. P.A.U., Ludhiana.

Schoonbee, H.L., Nakani. V.S. and Prinsloo, J. (1979). The use of cattle manure and supplementary feeding in growth studies of the Chinese Silver Carp in Transkri. South Africa. *S. Afr. J. Sci.*, 75 (ii) : 489-495.

Schroeder, G.L. (1980). Fish farming in manure loaded ponds. ICLARM-SEARCH *Conf. Agri. Aquaculture Fmg. Systems,* Manila. Phillipines. pp. 73-86.

Sehgal, H.S., Kaur. K. and Sehgal. G.S. (1992). Zooplankton response to biogas slurry in carp ponds. *Bioresource Technology* 41 : 111-116.

Yadav, N.K. and Garg. S.K. (1992) Relative efficacy of different doses of organic fertilizer and supplement feed utilization under intensive fish farming. *Bioresource Technology* 42 : 61-65.

CHAPTER 2

The Use of Probiotic Feed Supplement in Indian Major Carps When Challenged with *Aeromonas Hydrophila*

Partha Bandyopadhyay, Arup Tewary, Aditya K. Dolai and Bidhan C. Patra[1]

Aquaculture Research Unit, Department of Zoology, Vidyasagar University, Midnapore - 721102, West Bengal, India

ABSTRACT

The present study is observed that probiotic feed supplement made a positive co-relation between immune response and survivality of Indian major carps; *Catla catla, Labeo rohita* and *Cirrhinus mrigala.* During the 60 days feeding trial locally available agro based ingredients were used with a commercial probiotic supplement. In case of blood parameters highest haemoglobin percentage, TLC, TEC were found significantly higher ($p < 0.05$) in *Cirrhinus mrigala* followed by *Catla catla* and *Labeo rohita.* After 10 days challenge trial, *Cirrhinus mrigala* showed highest survivality (40%) followed by *Labeo rohita* (35%) and *Catla catla* (20%) in case of probiotic feed supplement but all the fished were died before 10 days in case of control. From the above experiment it is clearly indicate that probiotic supplement feed gives more immune response than control and *Cirrhinus mrigala* has a better immune response than *Labeo rohita* and *Catla catla.*

Key words: Indian major carps, immune response, *Aeromonas hydrophila*

INTRODUCTION

Aquaculture is also growing faster than beef cattle production (10% vs 3%), Verschuere *et al.*, 2000. But, as with many other industries, this rapid growth has brought with it the

[1] Correspondent Author
Telephone : Office : (91) 03222-276554, Ext - 455 Residence : (91) 03222-266035
Fax No.: (91) 03222-275329 Email : bcpatra@yahoo.com

problem of disease outbreaks. Now, researchers are trying to use probiotic bacteria in aquaculture to improve water quality by balancing bacterial population in water and reducing pathogenic bacterial load. "Probiotics" literally means "for life" and an antonym to antibiotics, which involves multiplying few useful microbes to compete with the harmful ones, thus suppressing their growth (Mishra *et al.,* 2001). The therapeutic use of antibiotics result in intestinal disorders. This is believed to be due to the different effects of antibiotics, which tends to remove indigenous gut flora along with disease causing agents. This situation is avoided by an alternative situation in the production system through the use of beneficial bacteria to fight against pathogenic bacteria i.e., probiotics that is an acceptable practice in aquaculture. The health of the animal thus improved by the elimination of pathogens or at least minimizing the effects, which benefits aquaculture. Research network in this field of aquaculture will greatly help to intensify fish production in the country. An initiation has been made in this regard at Aquaculture Research Unit, Department of Zoology, Vidyasagar University, Midnapore, West Bengal, India, which needs to be further strengthened with specific and technical expertise drawn from various Research Organisations and Universities. Such effort will go a long way to evolve effective mechanism for providing sustainable aquaculture in the country.

MATERIALS AND METHODS

The present study was conducted partly at Vidyasagar University, Midnapore (Latitude and Longitude of this area 22° 25' N and 87° 20' E), and West Bengal, India. The Indian major carps, *Catla catla, Labeo rohita* and *Cirrhinus mrigala* were selected for the present investigation. Fishes were brought to the laboratory, acclimatised and released into glass aquaria (90 X 45 X 45cm^3; 182.251 capacity).

Each feed item was analysed for moisture, dry matter (DM), crude protein (CP) (Kjeldahl N x 6.25), crude lipid (CL) (ether extract) and ash as per Weende proximate analyses scheme following the methods outlined in the "Official Methods of Analysis" published by Association of Official Agricultural Chemists (AOAC, 1990).

To determine the feed conversion ratio (FCR) of the formulated feeds, the feeding experiment was continued for a period of 60 day. Each experiment was tried in duplicate. The fishes were fed daily @ 5% of their total body weight. The water of each aquarium was changed daily. The amount of feed given was adjusted on the basis of the weight gain of the sampled fish. The absolute conversion rate was then calculated as:

$$\text{Absolute conversion rate} = \frac{\text{Dry weight of the feed given}}{\text{Increase in wet weight of fish}}$$

PER of the formulated diets was estimated by the gain in wet weight of the fish per gram of crude protein consumed on dry weight basis:

$$\text{PER} = \frac{\text{Wet weight gain of the fish}}{\text{Gram of crude protein consumed}}$$

Gross energy values of the above were determined using 1341 parr pxygen bomb calorimeter. To record the increment in weight (g) and to calculate the specific growth rate %, Live weight gain (%) and daily weight gain (g) the following formula has been applied:

$$\text{Specific growth rate (\%)} = \frac{\text{Ln final weight - Ln initial weight}}{\text{Days on trial}} \times 100$$

$$\text{Live weight gain (\%)} = \frac{\text{Final weight - Initial weight}}{\text{Initial weight}} \times 100$$

$$\text{Daily weight gam} = \frac{\text{Wet weight gain}}{\text{Days on trial}}$$

Blood samples were collected in the field or in the laboratory for serological diagnosis by syringe from caudal vein and heart. Blood samples were collected by heparinized syringe from caudal vein for haematology. EDTA (Ethyline diamine tetraacetic acid) was used as anticoagulant. 1.0 mg EDTA ml^{-1} of blood or 1 drop of 1.0% solution 5 ml^{-1} of blood was used for haematology. The Gornall's biuret method (Kulow, 1967) has proved to be particularly suitable for the routine investigation of fish. Fractionation of the serum proteins was done with the 22.6% sodium sulfate solution (22.6 g. of anhydrous Na_2SCO_4 in 100 ml of distilled water). 0.5 ml of serum was mixed with 9.5 ml of Na_2SO_4 solution in the centrifuge tube and shook end-over-end. 2.0 ml of solution was removed and 3.0 ml of ether was added to the remaining 8.0 ml of the mixture. Centrifuge tube was closed with the stopper, then shook for 30 seconds and centrifuged for five minutes. 2.0 ml of the removed mixture (mixture-A) was used to determine the total proteins, following Gornall. The other part of the mixture (mixture-B) was used to determine albumins, following Gornall.

The globulin content is the difference between the total proteins and the albumins.

Globulins = Total proteins — Albumins.

With the help of the albumin and globulin fractions, the albumin-globulin ratio can now be obtained, which can contribute to an assessment of the condition.

Dacie's fluid (Formaldehyde - 10 ml., Trisodium Citrate - 31.3 g., Brilliant Cresy blue - 1.0g and Distilled Water - 11.) was used for Total Erythrocyte Count.

The blood was drawn up to 0.5 mark in RBC diluting pipette of haemocytometer. The pipette was immediately filled to the 101 mark with the diluting fluid. The pipette was shaken for 30 seconds and then few drops of diluting blood were expelled from it. The tip of the pipette was touched to Neubauer's slide and cover slip junction. The Neubauer's slide is divided into ruled areas of 1 sq.min with the center sq.mm divided into 25 groups of 16 small squares each. The cells within the boundaries of five of these small squares (80 smallest squares) were counted.

So, TEC = total number of cells in five small squares x 10,000 cu.mm^{-1} of blood. Shaw's solution (Solution A - Neutral red - 25 mg., Sodium chloride 0.9 g, Distilled water 100 ml; Solution B - Crystal Violet 12.0 mg, Sodium Citrate 3.8 g, Formaldehyde 0.4 ml, Distilled water 100 ml) was used after filtering. Blood was drawn in RBC pipette up to 0.5 mark, then solution A up to half portion of blood and solution B up to 101 mark.

The four large squares were counted.

So, TLC = total number of cells in four large squares x 500 cu.mm^{-1} of blood.

Various methods are in use of which cyanomethaemoglobin method is most widely recommended. Cyanomethaemoglobin method is a colorimetric or spectrophotometric procedure for determining haemoglobin concentration. An aliquot of well-mixed whole blood is taken and reacted with a solution of potassium cyanide and potassium ferricyanide (Drabkin's solution). The chemical reaction yields a product of stable colour - the cyanomethaemoglobin. The intensity of the colour is proportional to the haemoglobin concentration and obeys Beer's law.

5 ml of Drabkin's solution [Potassium cyanide (KCN) - 50 mg, Potassium ferricyanide ($KFeCN_6$) - 200 mg, Distilled water - 1000 1] was taken in a test tube and then an aliguot (20 μ1) of well-mixed EDTA - anticoagulated blood specimen blowed into the Drabkin's solution. After 10 minutes absorbance was taken at 540 nm.

$$\text{Concentration of Haemoglobin in test solution (g dl}^{-1}\text{)} = \frac{Ax64458xF}{44xdx1000} x\ 10$$

[Where A = Absorbance of the solution at 540 nm.

64458 = Approximate molecular weight of Hb.

F = Dilution factor = 5/0.02 (20ml = 0.020 ml) = 250

d = Distance of light path in centimeter, which is always 1 cm.

1000 = Conversion of mg to g.

10 = conversion factor for g L^{-1} to g dl^{-1}]

In simple term, the above equation boils down to concentration of Hb

(g dl^{-1}) = A_{540} x 36.6.

[Note — The extinction co-efficient is expressed in millimole where the total volume is in liter. Hence, to convert mg to g divided by 1000 and to convert liter (1000 ml) to deciliter (100 ml) divide by 10].

10 days challenge trial was done with *Aeromonas hydorphilla* treated with different experimental diets of Indian major carps.

Duncan multiple range tests were used for statistical analysis (Alder and Roessler, 1972). The linear regression equation, correlation between control and experimental diets were analysed for the best fit curve and to study the rate of increment by using the software package like Microsoft Excel, STATISTICA, SPSS and ASP etc.

RESULT

After that challenge trial was conducted with *Aeromonas hydorphilla* for 10 days. The result compile various finding during the study under the following headings.

The proximate composition of ingredients (mustard oil cake, rice polish and fish meal) has been presented in Table - 2.1. The table indicates moisture content, dry matter, crude protein, crude lipid, ash content and energy levels. The table showed the crude protein was highest in fish meal. Being 48.65 ± 0.33% while the lowest was in rice polish 13.03 ± 0.17%. The crude lipid content was highest in mustard oil cake 12.24 ± 0.06% followed by fish meal and rice polish.

Table 2.1 Proximate composition of ingredients

Ingredients	Proximate composition					
	Moisture Matter	Dry Protein	Crude Protein	Crude Lipid	Ash	Energy
Mustard oilcake	6.05 ± 0.56[a]	93.95 ± 1.62[a]	39.23 ±0.11[a]	12.24 ± 0.06[a]	9.55 ± 0.09[a]	4.764 ± 0.08[a]
Rice polish	4.38 ± 0.32[b]	95.62 ±2.83[a]	13.03 ±0.17[b]	5.14 ±0.21[b]	21.41 ± 0.06[b]	3.519 ± 0.03[b]
Fish meal	2 30 ±0.19[C]	97.70 ±0.87[a]	4865 ±0.33[C]	672 ±0.14[C]	11 98 ±0.19[C]	4627 ± 0.20[a]

Expressed as percentage of fresh weight, while crude protein, crude lipid, Ash and energy are expressed as percentage of dry matter.

Results are means of five separate determinations (Mean ± SD)

Figures having different alphabets (superscribed) in the same column are significantly different ($p < 0.05$)

Feed formulation of different experimental diets are presented in Table - 2.2. All the formulated feeds are isocaloric and isonitrogenus. LC, CC and MC were control feed i.e. with without probiotic supplement while LP, CP and MP were containing 0.5% probiotic supplement in replacing of tapioca floor.

Table 2.2. Formulation of different experimental diets* for Indian major carps

Ingredients	Compositic (% DM)					
	LP	LC	CP	CC	MP	MC
1	2	3	4	5	6	7
Mustard oil cake	40	40	40	40	40	40
Rice polish	20	20	20	20	20	20

1	2	3	4	5	6	7
Fish meal	30	30	30	30	30	30
Vitamin and Minaral mixture*	2	2	2	2	2	2
Cod liver oil**	2	2	2	2	2	2
Tapioca	5.5	6.0	5.5	6.0	5.5	6.0
Probiotics***	0.5	00	0.5	00	0.5	00

*Vitamin minaral mixture: (KALVIMIN FORTE) composition of vitamin: Vitamin-A, 50,00,000 IU. Vitamin-B2, 2.0 gms. Vitamin-B12, 6.0 mg. Vitamin-D3, 10,00,000 IU. Cal. Pantotherate-4.0 gm. Composition of Minaral: Calcium-800 gm. Phosphorus-150gms. Manganese-27.5 gms. Iodine-1.0 gms. Iron-7.5gms. Zinc-15.0 gms. Copper-2.0 gms.

**Cod liver oil (SEACOD, M/s Universal Medicare Ltd., Mumbai) rich in Vitamins A & D and other health-promoting nutrients like EPA and DHA

***Probiotics (SPILAC) : Each kg. Contains- Lactobacillus sporogenes-33,200 cfu. Liver extract-5320mg. Yeast extract *(Saccharomyces cervisiae)* 1250mg. Alpha amylase-5000 mg. SPIRULINA-5320mg. Ascorbic acid coated-500 mg. Xanthophyll-9000 ug. Chlorophyll-9500 ug. GLA-38000 ug. Protein-100 gm.

The proximate composition of the diets (LP, LC, CP, CC, MP and MC) have been presented in Table - 2.3. The percent of crude protein, crude lipid, ash and energy values are not significantly different ($p < 0.05$) between the diets.

Table 2.3 Proximate composition of different experimental diets* for Indian major carps

Proximate composition of feeds	Diets					
	LP	LC	CP	CC	MP	MC
Crude protein	32.88 ±0.13^a	32.87 ±0.22^a	32.87 ±0.32^a	32.82 ±0.32^a	32.78 ±0.32^a	32.86 ±0.29^a
Crude lipid	7.91 ±0.07^a	7.92 ±0.09^a	7.91 ±0.11^a	7.94 ±0.23^a	7.89 ±0.31^a	7.90 ±0.09^a
Ash	11.15 ±0.16^a	11.13 ±0.08^a	11.05 ±0.06^a	11.18 ±0.19^a	11.16 ±0.07^a	11.15 ±0.10^a
Energy (Kcal/gm)	394.95 ±0.42^a	394.08 ±0.27^a	394.35 ±0.33^a	394.62 ±0.38^a	393.98 ±0.87^a	394.10 ±0.20^a

*On dry matter basis.

Results are means of five separate determinations (Mean + SD)

Figures having same alphabets in the different row are significantly different ($p < 0.05$)

The variation in the level of crude protein, crude lipid and gross energy for initial and final body carcass have been presented in Table - 2.4.

Table 2.4 Initial and Final carcass composition of Indian major carps of the 60 days experimental trial

Carcass Composition	Experimental tanks						
	Initial	Final					
		LP	LC	CP	CC	MP	MC
Crude protein	60.26 ± 2.97^{a}	63.24 ± 2.94^{b}	61.02 ±3.01^{a}	64.26 ±2.21^{b}	62.01 ± 2.92^{a}	61.83 ± 3.03^{a}	60.92 ± 2.98^{a}
Crude lipid	16.24 ± 1.71^{a}	20.51 ±1.31^{b}	19.88 ±2.12^{b}	22.58 ±2.93^{b}	21.89 ± 1.29^{b}	19.80 ±1.17^{b}	18.20 ± 1.38^{b}
Energy (kJ/g)	16.01 ±1.12^{a}	20.09 ±1.31^{b}	19.90 ±2.13^{C}	20.95 ±1.18^{b}	19.83 ±1.31^{b}	20.02 ± 2.24^{b}	18.80 ±1.39^{b}

Results are means of five separate determinations (Mean ± SD).

Figures having different alphabets (superscribed) in the same row are significantly different ($p < 0.05$)

The growth of Indian major carps in relation to various diets are presented in figure - 1.1. The highest net weight gain (NWG) was observed in case of CP 225.93 ± 1.36 gm followed by MP, CC, MC, LP and LC, while lowest NWG (94.93 ± 1.21g) was observed in case of LC which is a control feed (Table - 2.5).

Table 2.5 Net weigh gain, Weigh gain (%), FCR, SGR and PER of Indian major carps in relation to various diets

Diets	NWG	ADG	SGR	FCR	PER
LP	132.50 ± 1.29^{a}	2.20 ± 0.002^{a}	2.83 ± 0.30^{a}	4.31 ±0.18^{a}	6.20 ±0.14^{a}
LC	94.93 ±1.21^{b}	1.58 ±0.001^{b}	2.95 ±0.41^{a}	5.40 ±0.33^{b}	4.44 ±0.11^{b}
CP	225.93 ±1.36^{C}	3.76 ± 0.007^{C}	3.26 ± 0.39^{b}	4.52 ± 0.34^{a}	10.57 ± 0.32^{C}
CC	184.40 ± 1.96^{d}	3.07 ±0.011^{d}	2.83 ± 0.30^{a}	3.80 ± 0.44^{o}	8.62 ±0.19^{d}
MP	195.07 ±2.11^{e}	3.25 ±0.016^{d}	2.43 ±0.21^{C}	2.99 ± 0.28^{d}	9.12 ± 0.23^{e}
MC	140.54 ± 0.95^{f}	2.34 ± 0.003^{3}	2.44 ±0.18^{C}	3.42 ± 0.56^{e}	8.57 ± 0.27^{d}

Results are means of five separate determinations (Mean ± SD)

Each replicate contains 10 samples.

Figures having different alphabets (superscribed) in the same column are significantly different ($p \leq 0.05$)

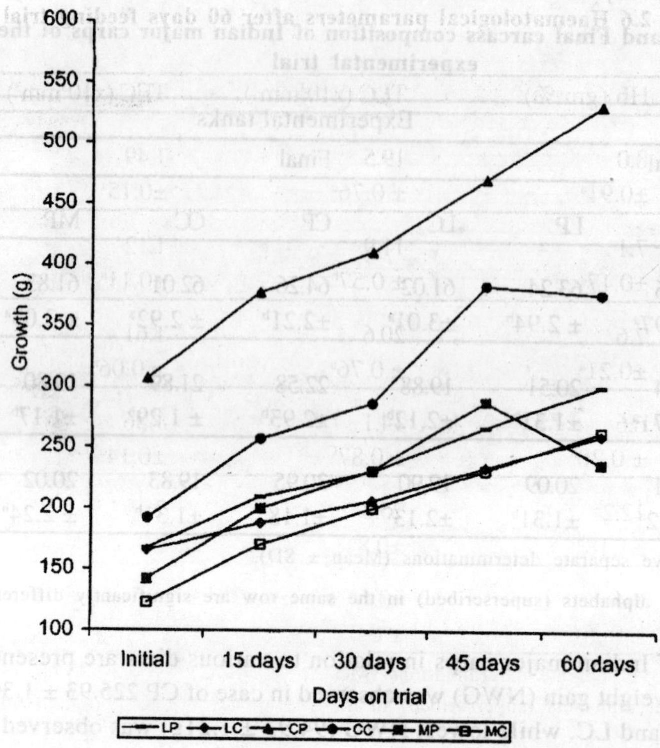

Figure 2.1 : The growth of Indian major carps in relation to various diets.

The average daily growth, FCR, SGR and PER is represented in Table - 2.5. The specific growth rate (SGR) showed the highest value (3.26 ± 0.39) while lowest was observed in MP 2.43 ± 0.21.With regard to feed conversion ratio (FCR) during the study period LC had showed highest (5.40 ± 0.33) while lowest in MP (2.99 ± 0.28). Protein efficiency ratio (PER) was highest in CP (10.57 ± 0.32) followed by MP (9.12 ± 0.23), CC (8.62 ± 0.19), MC (8.57 ± 0.27), LP (6.20 ± 0.140) and LC (4.44 ± 0.11) during the study period.

The variation in the level of Hb, TLC and TEC have been represented in Table - 2.6. Haemoglobin (Hb) percentage was not significantly ($p < 0.05$) different between the test animals. But in case of total lucocyte, highest was observed in MP (25.8 ± 0.81 10^3mm^3) significantly ($p < 0.05$). Similar trend was found in case of total erythrocyte count also. The variation in the level of total serum protein, albumin, globulin have been represented in Table - 2.7. Highest total serum protein (TSP) was observed in MP (3.61± 0.49 g%) and lowest in MC (2.33 ± 0.13 g%). Highest globulin g% was observed in MP (2.73 ± 0.25) followed by MC (2.50 ± 0.21), LP (2.35 ± 0.44), CP (2.14 ± 0.20) LC (2.06 ± 0.03) and CC (1.87 ± 0.09). The highest albumin (g%) was observed in LP (0.91 ± 0.04) while lowest in CP (0.68 ± 0.03).

Challenge trial was done with *Aeromonas hydorphilla* treated with different experimental diets of Indian major carps have been presented in Table - 2.8. Data indicated that 100% mortality was observed in CC in 5th day followed by LC in 6th and MC in 8th day. Highest survivility was observed in MP (40%) followed by LP (35%) and CP (20%) on the 10th day of the challenge trial.

Table 2.6 Haematological parameters after 60 days feeding trial

Treatment	Hb (gm %)	TLC ($x10^3mm^3$)	TEC ($x10^3mm^3$)
LP	8.0 ±0.91[a]	19.5 ± 0.76[a]	1.49 ±0.15[a]
LC	7.1 ±0.17[a]	11.9 ± 0.57[b]	1.12 ±0.11[a]
CP	7.6 ±0.21[a]	20.6 ± 0.76[a]	1.61 ± 0.06[a]
CC	6.6 ± 0.20[a]	14.1 ± 0.87[C]	1.36 ±0.14[a]
MP	12.2 ±1.18[b]	25.8 ±0.81[d]	2.19 ±0.21[b]
MC	9.4 ± 0.20[a]	10.4 ± 0.32[b]	1.88 ± 0.09[a]

Results are means of five separate determinations (Mean ± SD)

Figures having different alphabets (superscribed) in the same column are significantly different (p≤0.05)

Table 2.7. Serological parameters after 60 days feeding trial

Treatment	TSP (gm%)	Albumin (gm%)	Globulin (gm%)
LP	3.26 ± 0.44[b]	0.91 ± 0.04[a]	2.35 ± 0.44[a]
LC	2.80 ±0.18[a]	0.74 ± 0.02[b]	2.06 ± 0.03[a]
CP	2.82 ±0.19[a]	0.68 ± 0.03[C]	2.14 ± 0.20[a]
CC	2.65 ±0.14[a]	0.78 ± 0.03[b]	1.87 ± 0.09[a]
MP	3.61 ± 0.49[b]	0.88 ± 0.02[a]	2.73 ±0.25[a]
MC	2.33 ±0.13[a]	0.17 ±0.01[d]	2.50 ±0.21[a]

Results are means of five separate determinations (Mean ± SD)

Figures having different alphabets (superscribed) in the same column are significantly different (p≤0.05)

Table 2.8. Challenge trial with *Aeromonas hydrophila* treated with different experimental diets of Indian major carps

Treatment	Mortality % in challenged experimental fishes										
	Days										
	1	2	3	4	5	6	7	8	9	10	Survivality
LP	0	0	5	10	20[a]	40	50	54	62	65	35
LC	0	0	15	50	80[b]	100	-	-	-	-	0
CP	0	0	10	25	40[c]	50	55	70	75	80	20
CC	0	0	20	60	100[d]	-	-	-	-	-	0
MP	0	0	0	0	10[e]	25	40	45	55	60	40
MC	0	0	5	30	45[f]	70	90	100	-	-	0

Figures having different alphabets (superscribed) in the same column are significantly different ($p \leq 0.05$)

DISCUSSION

Present study is observed that probiotic feed made a positive co-relation between growth survival dietary utilization and immuno response of Indian major carps. Similar observations was confirmed by Requelme *et al*, 1996 in case of oyster larvae, Ravi *et al.*, 1998 in case of *Peneaus indicus*, Gram *et al.*, 1999 in case of Rainbow trout, Nagami *et al.*, 1997 in case of blue crab also. Specific growth rate was highest in *Catla catla* in probiotic supplement and gives a positive co-relation with protein efficiency ratio, i.e. dietary protein intake and which was also observed by Tacon and Cowey 1985. Highest FCR values were observed in case of control feed and lower in probiotic supplement feed indicates better nutrient utilization. So, net weight gain, average daily growth showed significantly higher values ($p < 0.05$) in Indian major carps fed probiotic feed. In a present study there was no mortality observed during the 60 days feeding trial because of control condition, irrespective of the level of protein of the prepared diets.

While comparing with a probiotic feed with control feed during the study it was observed that probiotic feed showed better result than the control feed. While the cost of probiotic more or less similar or insignificantly different ($p < 0.05$). It is important here to record that the probiotic feed used in the present study in viable for sustainable carp culture for better water stability also. In present study equal proportion of Tapioca as a best binder was added in all the formulated diets (Bandyopadhyay, 2000) and 10.3% leaching is acceptable for the feed till with consume by the fish. In present study all the formulated feed are less than 10.3% leaching after one hour as Indian major carps is the reflexive feeder. Water stability beyond one hour is appreciable.

In case of carcass composition CP fed *Catla catla* showed high percentage of crude protein 64.26 ± 2.21 and lipid 22.58 ± 2.93 where LP fed *Labeo rohita* and MP fed *Cirrhunus mrigala* showed insignificantly ($p < 0.05$) different crude protein and lipid percentage but in case of control. *Catla catla, Labeo rohita* and *Cirrhinus mrigala* showed significantly different ($p < 0.05$) than probiotic supplement feed. The beneficial effect of probiotic in larval rearing have been demonstrated by Charlon and Bergot (1984), Dabrowski and Kaushik (1985), Bergot *et al.*, (1986) and Alami - Durante *et al.*, (1991). Singh *et al.*, (1980) observed that the yeast-incorporated feed are more readily acceptable by the fry of rahu *(Labeo rohita).* Charlon and Bergot (1984) achieved more than 89% survival and good growth of carp larvae fed exclusively on dry diets having yeast powder and pork/beef freeze dried liver. Dabrowski *et al.*, (1984) recorded significant growth in coregonid larvae involving two sepatare yeast-based diets. Bergot *et al,* (1986) also achieved high survival (88 - 92%) and growth rates in coregonid larvae fed with yeast powder and beef liver supplemented with cod liver oil, vitamin minerals. Charlon and Bergot (1984) reported 87% survival and an FCR of 0.62 in common carp *(Cyprinus carpio)* larvae reared on liver, yeast and commercial trout starter feed. Alami-Durante *et al.*, (1991) reported the highest survival of 95% and growth of 189 mg in larvae of common carp fed on a yeast-based diet supplemented with vitamins and minerals each at 5% level in a 21-day experiment. Singh *et al.*, (1980) also found that yeast increased the rate of feed intake and conversion efficiency.

After 10 days challenge trial *Cirrhinus mrigala* showed highest survivality (40%) found by *Labeo rohita* (35%) and *Catla catla* (20%) in case of probiotic feed supplement (.5%). But in case of control, *Catla catla* were died (100%) in 5th day, *Labeo rohita* in 6th day and *Cirrhinus mrigala* in 8th day and it is very promising because *Aeromonas hydrophila* is most dangerous in case of Indian major carps (Bandyopadhyay *et al.*, 2003).

Mohanty *et al.*, (1993) reported (100%) survival using liver-based diet containing goat liver, probiotics, cod liver oil, starch, vitamin and mineral mixture with average growth of (133.3 mg/fry) in rahu reared from spawn to fry. Similar results were obtained with respect to growth (53 mg/fry) and survival (97%) in the case of catla *(Catla catla)* fed on the same diet (Mohanty *et al.*, 1993) Further, Swain *et al.*, (1994) reported that catla spawn reared for 20 days at a stocking density of 60 million per hectare have resulted in a higher average weight of 67 mg with 87% survival when provided with a liver-Biobost based diet.

In case of heamotological parameters highest Hb percentage was found in mrigal fingerling fed with probiotic supplement diet. TLC was found significantly ($p < 0.05$) higher (25.8 ± 0.81 10^3 mm^3) in case of MP, which denotes higher immuno response and better survivality. MP also showed higher percentage of TEC (2.19 ± 0.21 10^6 m^3) followed by LP and CP. In case of serological parameters highest globulin percentage was found in case of MP (2.73 ± 0.25) followed LP and CP.

From the above haematological and serological parameters, it is clearly indicates that probiotic supplement feed gives higher immuno response than control. It is also indicates that *Cirrhinus mrigala* gives more immuno response followed by *Labeo rohita* and *Catla catla.*

There was no disease observed during the experimental period. This may be due to the supplementation of bacterial spores, very well justified by Gatesoupe (1991) when the turbot *(Scophthalmus maximus)* larvae fed rotifers with probiotics (lactic acid bacreria and spores of *Bacillus toyot)* not only improved the growth rate of fish larvae but also probably released antibiotics, thereby not being affected by certain opportunistic *vibrios*. As this is a preliminary study on the use of probiotics, the exact dose of probiotics in the fish feed for different stages would be much rewarding.

ACKNOWLEDGEMENTS

The authors are highly thankful to the Head, Department of Zoology, Vidyasagar University, India for necessary facilities and the research scholars of the Aquaculture Research Unit for their valuable co-operation during preparation of this article and the University Central Library for providing us the Internet facilities.

REFERENCES

Alami-Durante, H., N. Charlon, A. M. Escaffre and P. Bergot, 1991. Supplementation of artificial diets for common carp *(Cyprinus carpio* L) larvae. Aquaculture, 93: 167-175.

AOAC, 1990. Official Methods of Analysis. *Association of Official Anaytical Chemist.* 15th Edn. Edited by S. Williams, AOAC, Virginia, 1298 pp.

Bandyopadhyay, P., 2000. Relative efficiency of formulated diets with different protein levels and a commercial feed on growth, survival and dietary utilization of gold fish *(Carassius auratus* Linn.). M. F. Sc. Dissertation. Barkatullah Univ. India, pp 123

Bandyopadhyay, P., Swain, S. K., Das, B. K. and Patra, B. C., 2003. Pathobiology of *Aeromonas hydrophila* infection in Gold fish *(Carassius auratus* Linn). Aquacult 4(1): 129- 134

Bergot, P., N. Charlon and H. Durante, 1986. The effect of compound diets feeding on growth and survival of coregonid larvae. Arch. Hydrobial. Beith. Ergepn. Limnol., 22: 265-275.

Charlon, N. and Bergot, 1984. Rearing system for feeding fish larvae on dry diets. Trial with carps *(Cyprinus carpio* L.) larvae. Aquaculture pp. : 1-9.

Dabrowski, K., and S.J. Kaushik, 1985. Rearing of coregoind *(Coregonus schinzi palea* Cuv. *et* Val.) larvae using dry and live food. III. Growth of fish and developmental characteristics related to nutrition. Aquaculture, 48: 123-135.

Dabrowski, K., N. Charlon, P. Bergot and S.J. Kaushik, 1984. Rearing of coregoind *(Coregonus schinzi palea* Cuv. *et* Val.) larvae using dry and live food. I. Preliminary data. Aquaculture, 41: 11-20.

Gatesoupe, F.J., 1991. The effect of three strains of lactic bacteria on the production rate of rotifers, *Brachionus plicatilis*, and their dietary value for larval turbot, *Scophthalmus maximus.* Aquaculture 96: 335-342.

Gram, L., Melchiorsen, J., Spanggaard, B., Huber, I. and Nielsen, T. F., 1999. Inhibition of *Vibrio anguillarum* by *Pseudomonas fluorescens* AH2, a possible probiotic treatment offish. Appl. Environ. Microbiol. 65: 969 - 973.

Kulow, H., 1967. Eine Schnellmethode zur Bestimmung der Serumproteine von Satzkarpfen (A rapid method of finding the serum proteins in young common carp). Dt. Fischerei-Ztg., 14 : 241 - 149.

Mishra, S., Mohanty, S., Pattnaik, P. and Ayyappan, A., 2001. Probiotics : possible application in aquaculture. Fishing Chimes, 21(1):31-37.

Mohanty, S. N., S. K. Swain and S. D. Tripathi, 1993. Growth and survival of rohu spawn fed on a liver based diet. J. Inland fish. Soc. India, 25(2): 41-45.

Nogami, K., Hamasaki, K., Maeda, M. and Hirayama, K., 1997. Biocontrol method in aquaculture for rearing the swimming crab larvae, *Portunus trituberculatus.* Hydrobiologia. 358 : 291 - 295.

Ravi, V., Khan, S. A., and Rajagopal, S., 1998. Influence on probiotics on growth of Indian white prawn, Penaeus indicus. J. Sci. Ind. Res., 57(10-11): 752 - 756.

Requelme, C., Araya, R., Uchida, A., Satomi, M. and Ishida, Y., 1996. Isolation of a native bacterial strain from the scallop *Argopecten purpuratus* with inhibitory effects against pathogenic vibros. J. Shellfish Res. 15 : 369 - 374.

Singh, B. N., V. R. P. Sinha and D. P. Charkraborty, 1980. Feed intake, absorption, conversion and growth of fry and fingerlings of rohu, Labeo rohita, Indian. Fish, 27: 193-200.

Swain, S. K., S. N. Mohanty and S. D. Tripathi, 1994. Growth and survival of catla spawn fed a dry artificial diet in relation to various stocking densities (Abstract). In: National Symposium on Aquacrops. Central Institute or Fisheries Education, Bombay, p. 55.

Tacon, A.G.J. and Cowey, J., 1985. Protein and amino acid requirements. In: P. Tytler and P. Calow (Eds.) Fish Energetics: New Perspectives. Croom Helm, London. 155-183.

Verschuere, L., Romabaut, G., Sorgeloss, P. and Verstraete, W., 2000. Probiotic bacteria as biological control agents in aquaculture. Microbiology and Molecular Biology Reviews. 64 (4): 655-671.

CHAPTER 3

EFFECTS OF MELATONIN ON THE MELANOPHORES OF FISH, *LABEO ROHITA* (HAM.)

Shashi Patil
School of Studies in Zoology, Jiwaji University, Gwalior - 474011, India

ABSTRACT

In fish, *Labeo rohita* adapted to black background intraperitoneal administration of melatonin (MEL) *(in vivo)* resulted in paling of the integument. In both, innervated and denervated melanophores *in vitro* MEL caused aggregation of the pigment. *In vitro*, studies revealed ineffectiveness of alpha (phenoxybenzamine and yohimbine) and beta (propranolol) adrenolytic agents to inhibit MEL induced responses. Adenosine (ADE), theophylline (THE) and alpha-melanocyte stimulating hormone (α-MSH) caused rapid dispersion of the pigment aggregated by MEL. It is suggested from the study that MEL induced rapid and reversible melanosome aggregation within melanophores of *Labeo rohita* and the receptors for its action are different from adrenoceptors. The action of MEL is assumed to be linked to inhibition of adenylate cyclase.

Key words : Melatonin, *Labeo rohita,* Aggregation, Dispersion, Receptors, Melanophores

INTRODUCTION

Lerner and his colleagues (1958) first discovered melatonin from mammalian pineal organ and named it after its potent concentrating action on amphibian melanophores. In many fish species melatonin has been reported to exert potent melanosome aggregation (Fujii and Oshima, 1986). while in others it does not influence melanosome translocation within melanophores (Fujii, 1993; Visconti and Castrucci, 1993). In restricted area of the skin the differential responsiveness to MEL exists (Fujii and Oshima, 1986; Masagaki and Fujii. 1999).

In fish, *Labeo rohita,* alpha adrenoceptors (Jain and Patil, 1992), beta adrenoceptors (Patil and Jain, 1991) and melanin concentrating hormone (MCH) receptors (Jain and Patil,

1990) exist. The present study was conducted to study the effects of melatonin on the melanophores of *Labeo rohita.*

MATERIALS AND METHODS

Specimen of fish, *Labeo rohita* of either sex weighing average 7g and measuring average 9 cm were acclimatized to laboratory conditions with normal day and light cycle at room temperature for few days before experimentation. Specimen were collected from Govt. Fish Farm, Gwalior (M.P.). For *in vitro* study the isolated scales were removed from antero-dorsal trunk region and immediately immersed in physiological saline (PS) for about 20 minutes for equilibration before treatment with drugs under study (Jain and Patil, 1992). Denervated melanophores were obtained at 2-6 hrs from fish injected with reserpine (10 μg/gi.p).

The responses of control and drug treated melanophores were recorded using melanophore index (M.I.) prepared for *Labeo rohita* (Patil and Jain, 1993) based on Hogben and Slome (1931) melanophore index. As many as 5 melanophores were observed from 5 to 7 fishes each for each drug. All values are expressed as mean ± S.D.M. The experiments were carried at room temperature - 20-22°C.

The macroscopic colour changes were recorded utilizing a method based on a 9 point scale derived from Munsell Grey series (Sharma et al., 1996).

The drugs/chemicals used are : Melatonin (Sigma, USA), α-melanocyte stimulating hormone (Peninsula Labs. Europe Ltd., U.K.), adenosine (Sigma, USA), theophylline (Sigma, USA), propranolol hydrochloride (Cipla, Bombay), phenoxy benzamine hydrochloride (Sigma, USA), yohimbine hydrochloride (Paul Neuendorf, Germany), reserpine (Hindustan Ciba Geigy Ltd., India). Working solutions of these drugs were prepared by diluting the stock solutions with PS immediately before use.

RESULTS AND DISCUSSION

A black adapted fish, *Labeo rohita,* with general tint of the body equivalent to index number (I.N. 2.5) paled notably within 5 min (I.N. 4.0) after receiving an intraperitoneal injection of MEL 10 mg/kg (Fig. 3.1). The maximum paling of skin was equivalent to I.N. 4.4 at 10 min that lasted for about an hour from the treatment. The influence of hormone declined (I.N. = 4.1 - 2 hr) but appeared to be a long lasting one, as even after 5 hr of its administration the fish were found to be appreciably pale (I.N. = 3.7). At 24 hr, on following day the pre-experimental shade was restored. This *in vivo* skin paling due to MEL is comparable with responses in other fishes (Bhargava and Jain, 1978; Prabhakar, 1988).

In vitro, melanophores were treated with MEL at concentrations ranging from 10^{-9}M to 10^{-5}M (Fig. 3.2). Minimal pigment aggregation was recorded at 10^{-8}M (M.I. 3.58, 10 min) and maximum at 10^{-5}M (M.I. 2.1, 10 min). The melanophores showed no further aggregation even at higher concentration. The aggregation responses in *Labeo rohita* are consistent with earlier reports for dermal melanophores in amphibians (Sugden, 1991; Rollag and Lynch, 1993) and many fish species (Fujii and Miyashita, 1978; Iwakiri, 1983 and Fujii and Oshima, 1986). In denervated melanophores also the MEL produced melanosome aggregation as in innervated

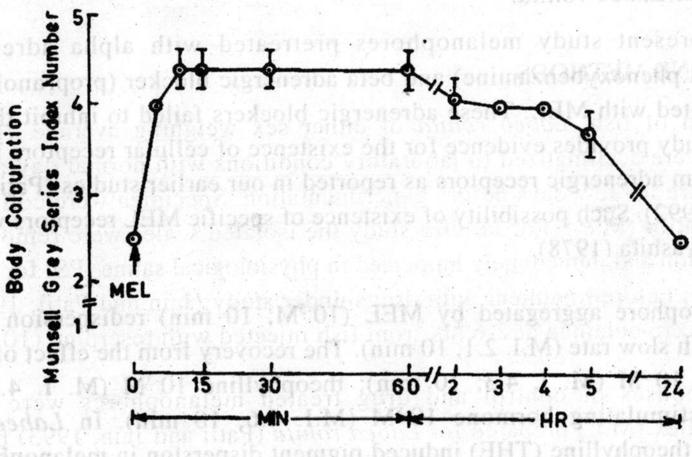

Fig. 3.1

In black-adpated fish, *Labeo rohita* placed over an illuminated black- background, a single injection of MEL (10 mg/kg) produced paling of the skin. Each point and the vertical bar represent the mean and ± S.D. (N = 5). respectively.

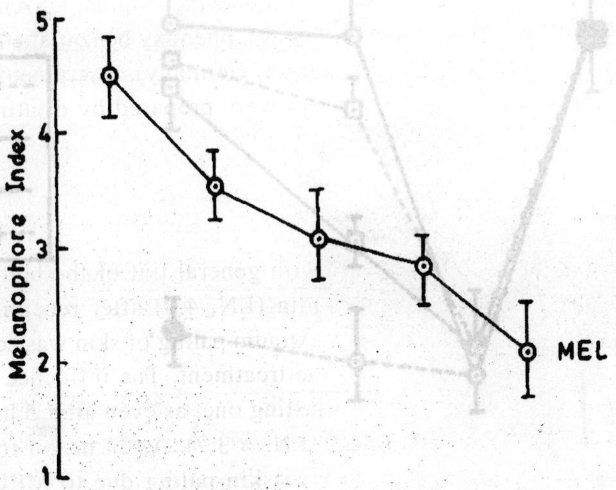

Fig. 3.2

Concentration-response curve of melatonin on isolated scale melanophores of *Labeo rohita.* Values are mean ± S.D. (vertical lines) of 25 measurements from 5 different fishes.

melanophores. Iwakiri (1983) reported these responses due to MEL in both innervated and denervated melanophores in fish, *Oryzias latipes.*

In the present study melanophores pretreated with alpha adrenergic blockers (Yohimbine and phenoxybenzamine) and beta adrenergic blocker (propranolol) at 10^{-4}M and 10^{-5}M were treated with MEL. These adrenergic blockers failed to inhibit the responses due to MEL. The study provides evidence for the existence of cellular receptors for the MEL that are different from adrenergic receptors as reported in our earlier studies (Patil and Jain, 1989; Jain and Patil, 1992). Such possibility of existence of specific MEL receptors was also reported by Fujii and Miyashita (1978).

In melanophore aggregated by MEL (10^{-5}M; 10 min) redispersion occurred in PS (control) but with slow rate (M.I. 2.1, 10 min). The recovery from the effect of MEL was rapid with adenosine 10^{-2}M (M. I. 4.3; 10 min); theophylline 10^{-3}M (M. I. 4.1; 10 min) and α-melanocyte stimulating hormone 10^{-8}M (M.I. 4.6; 10 min). In *Labeo* melanophores methylxanthine theophylline (THE) induced pigment dispersion in melanophores aggregated by MEL thus counteracting its action (Fig. 3.3). The effects of methylxanthine are in accordance with earlier studies on other fish species (Fujii and Miyashita, 1976 and Nagahama et al., 1980).

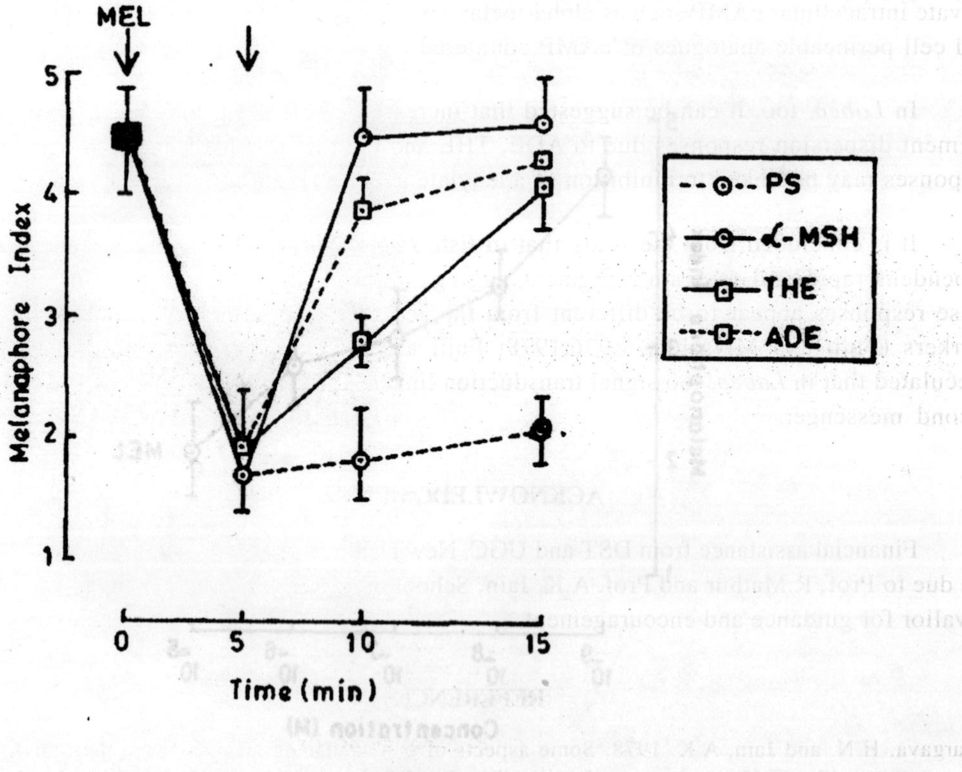

Fig. 3.3

Effects of ADE (10^{-2}M), THE (10^{-3}M) and α-MSH (10^{-8}M) on aggregatory responses of *Labeo* melanophores to MEL. Values are mean ± S.D. (vertical lines) from measurements on 5 different fishes.

Butcher and Sutherland (1962) explained it to be the result of increased level of cAMP due to methylxanthines. These phosphodiesterase inhibitors when applied externally they penetrate the cell along the large chemical gradient and lower the activity of cyclic nucleotide phosphodiesterase resulting in increased cAMP level and hence causing dispersion of the pigment.

Adenosine also dispersed the melanosomes within *Labeo* melanophores (Fig. 3.3). Fujii and Miyashita (1976) first suggested the role of adenosine and adenine nucleotides in the control of pigment dispersal in fish chromatophores. Several other studies (Oshima, 1989 and Namoto, 1992) support this pigment dispersing action of adenosine. In *Labeo rohita* there is evidence for the existence of adenosine receptors (unpublished observations) of A_2 type. These A_2-receptors stimulate the activity of adenylate cyclase (Londos et al., 1980 and Daly, 1983) elevating the cAMP level leading to dispersion of the pigment.

In the present study α-MSH mimicked the action of adenosine. It caused centrifugal translocation of the pigment in melanophores aggregated by MEL. The responses due to α-MSH are consistent with observations reported in other fishes (Sharma et al., 1996; Visconti et al., 1999 and Fujii, 2000).

In amphibian melanophores Seldenrijk et al. (1979) showed that the agents which elevate intracellular cAMP such as alpha-melanocyte stimulating hormone (a-MSH), forskolin and cell permeable analogues of cAMP countered the aggregation response of MEL.

In *Labeo,* too, it can be suggested that increased cAMP level may be responsible for pigment dispersion responses due to ADE, THE and α-MSH and MEL induced aggregation responses may be linked to inhibition of adenylate cyclase leading to decreased cAMP level.

It is concluded from the study that in fish, *Labeo rohita,* MEL induced concentration dependent rapid and reversible pigment aggregation and the cellular receptors mediating these responses appear to be different from the adrenergic receptors. As reported by many workers (Fujii and Miyashita, 1976,1978; Fujii and Oshima, 1986 and Fujii, 1993) it is speculated that in *Labeo,* too signal transduction linked to MEL action may involve cAMP as second messenger.

ACKNOWLEDGMENTS

Financial assistance from DST and UGC, New Delhi, is gratefully acknowledged. Thanks are due to Prof. R.Mathur and Prof. A.K. Jain, School of Studies in Zoology, Jiwaji University, Gwalior for guidance and encouragement.

REFERENCES

Bhargava, H.N. and Jain, A.K. 1978. Some aspects of colour change mechanism in the Indian fresh-water siluroid Heteropneustes fossilis (Bloch.). Biochem. Expt. Biol. XIV 4 : 359-373.

Butcher, R.W. and Sutherland, E.W. 1962. Adenosine 3'-5'-phosphate in biological materials. I. purification and properties of 3'-5' nucleotide phosphodiesterase and use of this enzyme to characterise adenosine 3'-5'-phosphate in human urine. J. Biol. Chem. 237: 1244-250

Daly, J.W. 1983'. Role of ATP and adenosine receptors in physiological processes. In physiology and pharmacology of adenosine derivatives (Eds. Daly, J.W., Kurioda, Y, Phillis, J.W.. Shimizu, H. and Ui, M.). Raven Press, New York. 275-290.

Fujii, R. 1993. Cytophysiology of fish chromatophores. Int. Rev. Cytol. 143 : 191-255.

Fujii, R. 2000. The regulation of motile activity in fish chromatophores. Pigment Cell Res. 13:300-319.

Fujii. R. and Miyashita, Y 1976. Beta-adrenoceptors, cyclic AMP and melanosome dispersion in guppy melanophores. Pigment Cell. 3 : 336-344.

Fujii, R. and Miyashita, Y. 1978. Receptor mechanisms in fish chromatophores-IV. Effects of melatonin and related substances on dermal and epidermal melanophores of the siluroid, *Parasilurus asotus.* Comp. Biochem. Physiol. 59C : 59-63.

Fujii, R. and Oshima, N. 1986. Control of chromatophore movements in teleost fishes. Zool. Sci. 3: 13-47.

Hogben, L.T and Slome, D. 1931. The pigmentary effector system. VI. The dual character of the endocrine co-ordination in amphibian colour change. Proc. R. Soc. London (Biol.). 108 B: 10-53.

Iwakiri. M. 1983. Action of melatonin and its derivatives on melanophores in isolated scales of a teleost fish. J. Sci. Hiroshima Univ. Ser B, Div I. 31 : 85-98.

Jain, A.K. and Patil, S. 1990. Evidence for melanin concentrating hormone (MCH) receptors mediating melanosome aggregation in *Labeo* melanophores Ind. J. Physiol. Pharmacol. 34 : 187-190.

Jain, A.K. and Patil, S. 1992. Alpha$_2$-adrenergic receptor activation induced melanophore response in a fresh-water teleost, *Labeo rohita :* an *in vitro* and *in vivo* study. Proc. Nat. Acad. Sci. India. 62B : 323-332.

Londos. C., Cooper, D.M.F. and Wolf, J. 1980. Subclasses of external adenosine receptors. J. Proc. Nat. Acad. Sci. USA. 77*: 2551-2554.

Lerner, A.B.; Case, J.D.; Takahashi, Y; Lee. T.H. and Mori, W. 1958. Isolation of melatonin the pineal gland factor that lightens melanocytes. J. Am. Chem. Soc. 80 : 2587.

Masagaki, A. and Fujii, R. 1999. Differential actions of melatonin on melanophores of the threeline pencilfish, *Nannostomus trifasciatus.* Zool. Sci. 16 : 35-42.

Nagahama, H.: Ohsaki, M. and Muneoka, Y. 1980. Action of methylxanthines on melanophores in isolated scales of a fresh-water teleost. J. Sci. Hiroshima Univ. Ser B Div 1. 28: 27-49.

Namoto, S 1992. Effects of purine compounds and forskolin on melanophores of the medaka *Oryzias latipes.* Evidence for A$_2$-adenosine receptor. Comp. Biochem. Physiol. 103C2:391-398.

Oshima. N. 1989. Adenosine inhibits the release of neurotransmitters from melanosome aggregating nerves offish. Comp. Biochem. Physiol. 93 C (2) : 207-211.

Patil, S. and Jain, A.K. 1989. The sympathetic neuro-melanophore transmission in a fresh-water. Indian major carp, *Labeo rohita* (Ham.). Ind. J. Physiol. Pharmacol. 33 : 101-106.

Patil, S. and Jain, A.K. 1991. Evidence for the involvement of beta-adrenoceptors in darkening responses in the fish *Labeo rohita.* Proc. Nat. Acad. Sci. Letters, India. 14: 191-193.

Patil. S. and Jain, A.K. 1993. Role of calcium in melanosome aggregation within *Labeo* melanophores. J. Bio. Sci. 18 : 83-91.

Rollag, M.D. and Lynch, G.R. 1993. Melatonin-induced desensitization in amphibian melanophores. J. Exp. Zool. 265 : 488-495.

Prabhakar. R.M. 1988. Photopigmentary responses in intact and eyeless fish *Labeo gonius* (Ham.). M.Phil. Dissertation. Jiwaji University, Gwalior, M.P., India.

Seldenrijk, R.; Hup, D.R.W.; De Graan, P.N.E. and Van De Veerdonk, F.C.G. 1979. Morphological and physiological aspects of melanophores in primary culture from tadpole of *Xenopus laevis.* Cell Tissue Res. 198 : 397-409.

Sharma, B.; Patil, S. and Jain, A.K. 1996. Effect of melanocyte stimulating hormone (MSH) on melanophores in isolated scales and on the colour displayed by the skin *(in vivo)* in a fresh-water teleost, *Puntius sophore* (Ham.). Proc. Nat. Acad. Sci. India. 6.6B : 139-150.

Sugden, D. 1991. Aggregation of pigment granules in single cultured *Xenopus laevis* melanophores by melatonin analogues. Br. J. Pharmacol. 104 : 922-927.

Visconti, M.A. and Castrucci, A.M.L. 1993. Melanotropin receptors in the cartilagenous fish *Petamotrygon reticulatus* and in the lung fish, *Lepidosiren paradpxa* Comp. Biochem. Physiol. 106 C. 525-528.

Visconti, M.A.; Ramanzini, G.C., Camargo, C.R. and Castrucci, A.M.L. 1999. Elasmobranch colour change : A short review and novel data on hormone regulation. J. *Exp.* Zool. 284 : 485-491.

CHAPTER 4

SELECTED HAEMATOLOGICAL PARAMETERS OF *CLARIAS BATRACHUS* AND *HETEROPNEUSTIS FOSSILIS* EXPERIMENTALLY INFECTED WITH *MYXOBACTERIA*

Prasad*, Yogendra and Verma, Vinay

**Aquatic Biot. & Fish Pathol. Lab.,*

Department of Animal Science, M.J.P. Rohilkhand University, Bareilly- 243 006 (INDIA)

ABSTRACT

Haematological investigations have been carried out on *C. batrachus* and *H. fossilis* experimentally infected with *Myxobacteria.* Inspite of the fact that the haematological values in fish are of clinical significance rather than commercial importance. Different blood parameters viz. TLC, DLC, TEC, Hb% and PCV% were examined and found to be highly significant differences in these parameters of artificially infected fish in comparison to asymptomatic and control fishes and followed the same pattern to that of naturally infected fish. This paper deals the impact of bacterial infection on the selected blood parameters of catfish.

INTRODUCTION

Farmed fish are inevitabaly subjected to various kinds of stresses that may lead to down regulation of immunity and as a result the out breaks of infectious diseases may occur. Bacteria are ubiquitously present in aquatic ecosystem and may be associated with the normal flora of fish skin while the pathogenic one causes diseases in fish that may be associated with the change in the different environmental conditions. These conditions include poor nutrition and water quality which may be conducive for the growth of pathogenic bacteria. Bacterial isolates which were found to be associated with symptomatic *C. batrachus* and *H. fossilis* and were identified as the members of gram negative bacteria *(Myxobacterium).*

Haematological parameters reflect the health status of fish (Conry, 1972) and affected by abiotic factors (temp., pH, DO, etc.) and biotic factors (infectious agents). They become highly active during microbial infections and lead to efficient clearance of pathogens (Dalmo *et al*, 1997). The main aim of this investigation was to findout the haematological changes arising in catfish *(C. batrachus* and *H. fossilis)* during the exposer of virulent *Myxobacterium* isolates. Moreover, limited informations are available on haematological parameters of Indian catfishes and few workers such as Prasad and Qureshi (1995), Mastan and Qureshi (2001), Kumari and Akela (2002) and Prasad and Verma (2003) have worked on the blood parameters of *Cirrhinus mrigala, Mystus cavasius, C. batrachus* and *C. batrachus* respectively subjected to Cytophagan or EUS infection.

MATERIALS AND METHODS

Indian catfishes viz. *C. batrachus* and *H. fossilis* were used as an experimental animals in the present study. Diseased fish were procured from the different water bodies with the help of fishermen and brought to the laboratory in living condition and kept in plastic pools (3'x2.5") filled with unchlorinated static bore well water. Different isolates *of Myxobacterium* were isolated from naturally infected *C. batrachus* and *H. fossilis* sampled from Nanak sagar and Sarda reservoirs.

BACTERIOLOGICAL EXAMINATION

Live and moribund symptomatic catfishes exhibited the symptoms of ulceration, necrotization, and haemorrhages on the general body surface were sacrified and their organs like kidney, heart, intestine and spleen were collected in saline solution (0.3% NaCl). Sampled organs were homogenized in the same solution and homogenates were centrifuged at 5000 rpm and supernatants were used for the isolation of bacteria on to Cytophaga agar (Anacker and Ordal, 1959) and *Myxobacterium* selective medium (Hsu *et al.*, 1983). The *Myxobacteria* inoculated plates were incubated at 25°C for about 48-72 hrs. The isolated strains were identified on the bases of different biochemical tests such as gram staining, shape, oxidase reaction, indole production, catalase production, gelatin degradation and H_2S production, citrate utilization, etc. following Cowan and Steel (1975).

EXPERIMENTAL INFECTION TRIALS

Virulence characterizations of Myxobacterial isolates were determined following the LD_{50}. For this purpose 0.2 ml. suspension of young colonies (48-72 hrs old) *of Myxobacterium* containing $2x10^5$, 2×10^6 and 2×10^7 cfu/ml were inoculated intramuscularly in three groups (five fish in each group) of healthy *C. batrachus* and *H. fossilis* and 50% mortality was recorded as per Reed and Muench (1938). Further, ten *C. batrachus* and ten *H. fossilis* were inoculated with 0.2 ml of bacterial suspension (containing $2x 10^5$ cfu/ml *Myxobacterium]* and kept in 500L plastic pools filled with clean, static and unchlorinated bore well water. Blood samples were collected from the lateral line of experimental fishes by inserting a heparinized microsyringe. The following methods were used for different blood parameters.

(a) Total Erythrocytes count (RBC, 10^6/mm^3)

The total erythrocytes count was estimated by the dilution chamber technique using Hayem's fluid as diluent (Benjamen, 1985). Blood samples were obtained as per standard technique. Briefly, the lateral line offish and the puncturing needle were sterilized with a small pad of cotton wool dipped into 90% ethanol. Thereafter lateral line was pricked with sterilized needle to get the blood. The first few drops of blood were wiped away and blood were sucked with R. B. C. pipette upto 0.5 mark. Excess blood, if any, was wiped away through the tip of the pipette and then Hayen's fluid was sucked up to 101 mark and mixed thoroughly to prevent haemolysis, coagulation and bacterial growth. Few drops of diluted blood were added under the coverslip of Neubauer's chamber and examined under microscope l0x. Five inner chambers (each chamber containing 16 squares), four at corners and one at centre were examined and the number of RBCs present in 16x5 squares were recorded. The total number of RBCs was calculated as per formula.

= Dilution x ^{1}/Volume number of cell counted (N)

= 200 x 50 x N

= 10,000 x N cells /mm^3.

(b) Total Leucocytes Count (W. B. C. 10^3/ mm^3)

Total leucocytes count was estimated by the standard dilution technique using diluting fluid (4% glacial acetic acid and two drops of gentian) (Talib and Khurana, 1995). Briefly, blood samples were obtained in the same way as described earlier and thereafter blood was sucked with W. B. C. pipette up to 0.5 mark. Excess blood, if any, was wiped away through the tip of pipette. Then diluting fluid was sucked upto 11 mark (1:20 dilution) and mixed thoroughly to prevent haemolysis, coagulation and bacterial growth. Few drops of diluted blood were added under the coverslip of Neubauer's chamber and examined under microscope in l0x. Four outside corners of the counting chamber were examined and the numbers of W. B. C. present in 16x4 squares were recorded. The total numbers of W. B. Cs. present were calculated as per formula.

The area of each large square is 1 sq. mm, the dept. of chamber being 0.1 mm and the volume of square is 0.1 m mm. volume of four corner squares = 0.1 x4 = 0.4 µ mm. No. of cells in four corner squares = N

0.4 µ m mm. contains = N cell

1 µ mm. contains = Nx20/0.4

= n50 = thousand/ mm^3.

(c) Different Leucocytes Count (DLC)

Differential count of blood smears prepared by staining with "Leishman stain" blue were made under 100X by using, oil immerson and counting was done by Meander method. Hundred cells were counted and tabulated viz. Neutrophils, Eosinophils, Basophils,

Lymphocytes andmonocytes. The total number of differential leucocytes present was calculated as per formula.

DLC = Number of types cells/Total number of white blood carpuscles x 100

(d) Haemoglobin (Hb%)

Haemoglobin was estimated by the shalis acid haematin method as per modified by Talib and Khurana (1995). N/10 HC1 solution (1.2 C. C. of conc. HC1 dissolved in 100 C. C. distilled water) was taken in graduated tube upto 2mm. The micropipette was filled by sucking the fresh blood upto the make of 20 mm. thereafter, small amount of blood adhering the micropipette was washed off by sterilized cotton. The blood of micropipette was now added to N/10 HC1 solution of graduated tube. Both the solutions were mixed by glass rod for 10 min. Thereafter, graduated tube was filled in Haldames haemoglobinometer (Haemometer) and distilled water was added drop by drop until the colour of the experimental tube solution become same as that of the standard tube. The level of experimental solution indicated the % of haemoglobin content.

(e) Packed Cell Volume (PCV)

Microhaematocrit values were estimated following the Benjamin (1985). Blood was drawn into a heparinized microhaemotocritic capillary tubes and their one end was seald with plasticine. It was placed in centrifuge and centrifuged at 10,000 rpm for 5 min. The capillary tubes were removed and reading were recorded.

RESULTS

Out of 2600 fish examined only 130 fishes were showing the symptoms of ulceration, necrotization and haemorrhages on the general body surface.

Haematological investigations have also been carried out by taking blood samples from control and artificially inoculated fishes and significant decline in the TEC, Hb, PCV and neutrophils was recorded which are presented in Tables-4.1 (a & b) and 4.2 (a & b).

Table-4.la : Impact of *Myxobacterium* on the different blood parameters of *Clarias batrachus*.

Fish group	TEC (10^6/mm^3)		TLC(10^3/mm^3)		Hb (%)		PCV (%)	
	normal	infected	normal	infected	normal	infected	normal	infected
1	2.75	1.9	2.2	2.98	13.0	9.2	36.7	27
2	2.97	1.7	1.78	2.48	15.4	9.4	45.4	33.2
3	2.95	2.15	1.84	2.55	13.8	8.6	38.9	30.5
4	2.85	1.6	1.97	2.64	14.4	9.2	44.5	32.4
5	2.87	1.8	2.01	2.81	14.2	8.7	39.5	32.8
	2.878	1.83	1.96	2.69	14.16	9.02	41.0	31.18
	±0.039	± 0.094	±0.073	± 0.090	±0.039	±0.185	±1.68	±1.14
sig.	0.000		0.000		0.000		0.001	

Table-4.1b : Effect *of Myxobacterium* on the DLC *of Clarias batrachus.*

Fish	Neutrophils(%)		Lymphocytes(%)		Monocytes(%)		Eosinophils(%)		Basophils(%)	
group	normal	infected	normal	infected	normal	infected	normal	infected	normal	infected
1	38	21	47	53	13	20	2	4	0	2
2	34	17	46	52	16	23	3	6	1	2
3	31	18	48	53	19	24	2	4	0	1
4	30	17	51	58	16	21	1	3	2	1
5	30	24	50	55	18	20	1	1	1	0
	32.6	19.4	48.4	54.2	16.4	21.6	1.8	3.6	0.8	1.2
	±1.539	±1.363	±0.927	±1.067	±1.029	±0.812	±0.374	±0.812	±0.374	±0.374
sig.	0.000		0.003		0.004		0.079		0.471	

Table-4.2a : Impact of *Myxobacterium* on the different blood parameters of *Heteropneustis fossilis.*

Fish group	TEC ($10^6/mm^3$)		TLC($10^3/mm^3$)		Hb (%)		PCV (%)	
	normal	infected	normal	infected	normal	infected	normal	infected
1	2.7	2.3	1.6	2.3	14.0	9.4	34.7	25.5
2	2.6	1.7	1.4	2.2	13.4	7.4	33.5	24.6
3	2.8	1.7	1.5	2.0	'12.6	8.6	33.0	25.8
4	2.5	1.8	1.7	2.5	14.6	8.8	33.8	23.7
5	2.9	2.2	1.4	2.1	12.8	9.5	34.6	27.5
	2.7	1.94	1.52	2.22	13.6	8.74	33.92	25.42
	±0.057	±0.180	±0.058	±0.086	±0.398	±0.388	±0.376	±0.636
sig.	0.003		0.000		0.000		0.000	

Table-4.2b : Effect *of Myxobacterium* on the DLC *of Heteropneustisfossilis.*

Fish group	Neutrophils(%) normal	Neutrophils(%) infected	Lymphocytes(%) normal	Lymphocytes(%) infected	Monocytes(%) normal	Monocytes(%) infected	Eosinophils(%) normal	Eosinophils(%) infected	Basophils(%) normal	Basophils(%) infected
1	34	15	50	60	13	17	2	5	1	3
2	33	17	51	58	14	18	2	6	0	1
3	34	14	49	61	14	16	2	7	1	2
4	33	13	52	60	13	19	1	6	1	2
5	35	15	51	59	13	17	1	6	0	3
	33.8	14.8	50.6	59.6	13.4	17.4	1.6	6.0	0.6	2.2
	±0.374	±0.663	±0.509	±0.509	±0.244	±0.509	±0.244	±0.316	±0.244	±0.374
sig.	0.000		0.000		0.000		0.000		0.007	

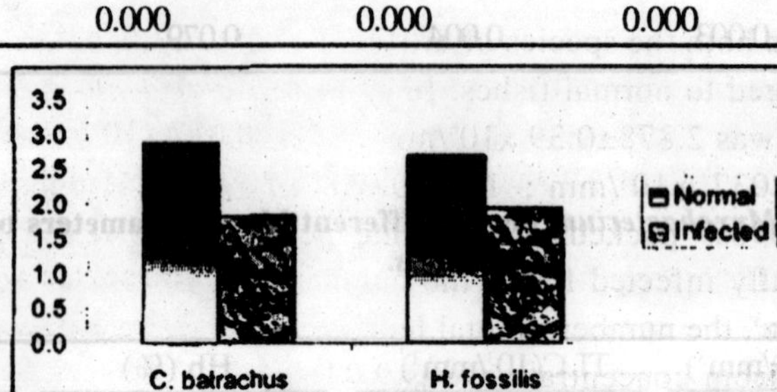

Fig. 4.1 Showing TEC of control and experimentally infected *C. batrachus & H. fossilis*

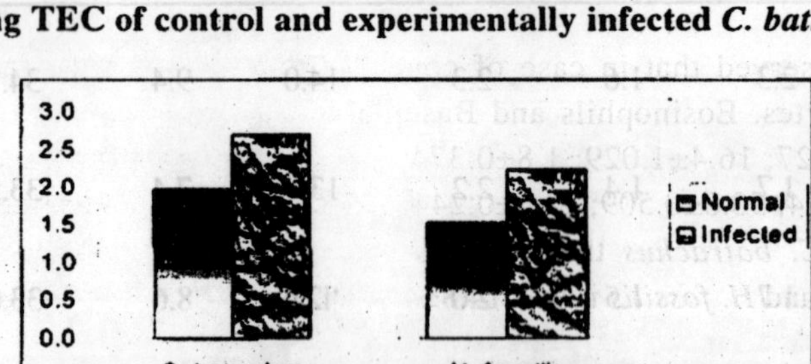

Fig. 4.2 Showing TEC of control and experimentally infected *C. batrachus & H. fossilis*

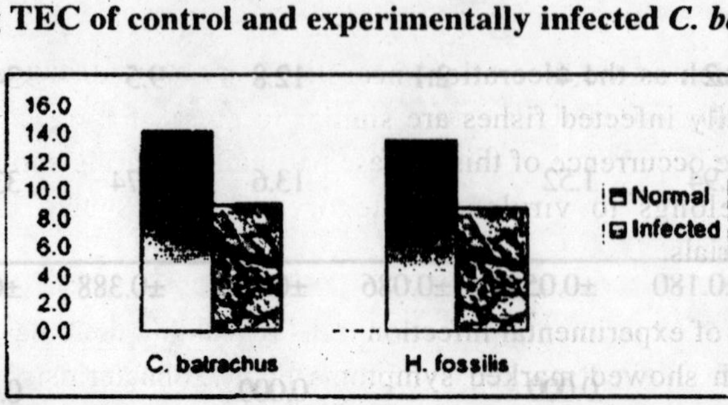

Fig. 4.3 Showing TEC of control and experimentally infected *C. batrachus & H. fossilis*

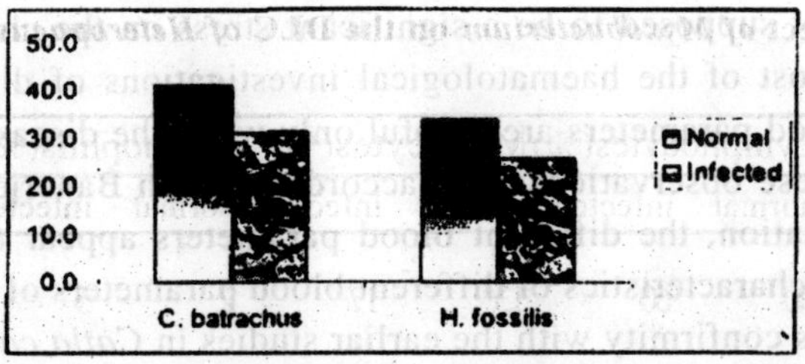

Fig. 4.4 Showing TEC of control and experimentally infected *C. batrachus* & *H. fossilis*

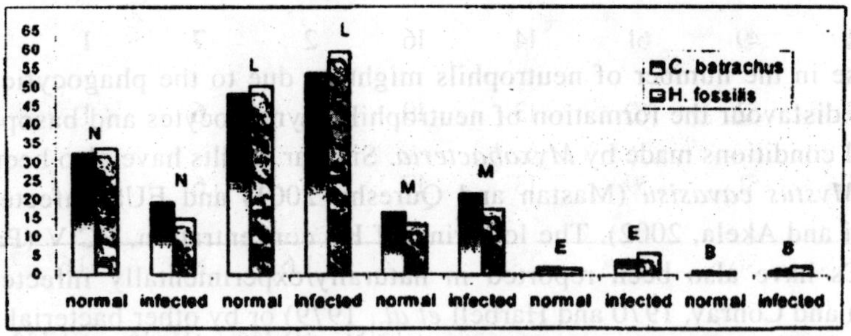

Fig. 4.5 Showing TEC of control and experimentally infected *C. batrachus* & *H. fossilis*

TEC was lower in both the species of artificially infected fishes, i.e. *Clarias batrachus* and *H.fossilis* as compared to normal fishes. In control *C. batrachus* and *H. fossilis* the total number of erythrocytes was 2.878±0.39 x10^6/mm^3, 2.70±0.057 x10^6/mm^3, the number of total leucocyte was 1.96±0.037 x10^3/mm^3; 1.52±0.058 x10^3/mm^3 Haemoglobin percent was 14.16±0.39%; 13.5±0.398% packed cell volume percent was 41.0±1 .68%; 33.92±0.376 respectively. In artificially infected fishes the number of erythrocytes was 1.83±0.094 x10^6/ mm^3, 1.94±0.18 x10^6/mm^3, the number of total leucocyte was 2.69±0.18 x10^3/mm^3, 2.22±0.086 x10^3/mm^3. The haemoglobin concentration was 9.02±0.185%; 8.74±0.338%; the packed cell volume was 31.18±1.14, 25.42±0.363 respectively.

It was also observed that in case of control group the mean values for neutrophils, Lymphocyte, Monocytes, Eosinophils and Basophils of a asymptomatic *C. batrachus* were 32.6±1.539; 48.4±0.927; 16.4±1.029; 1.8±0.374; 0.8±0.374 respectively and *H. fossilis* these values were 33.8±0.374; 50.6±0.509; 13.4±0.244; 1.6±0.244. 0.6±0.244 respectively. In case of artificially infected *C. batrachus* these values were 19.4±1.363; 54.2±1.067; 21.6±0.812; 3.6±0.812; 1.2±0.374 and *H. fossilis* it was 14.8±4.8±0.663; 59.6±0.509; 17.4±0.509; 6.0±0.316; 2.2±0.374.

DISCUSSION

The symptoms such as the ulceration, necrotization, haemorrhages and eroded mouth exhibited by the naturally infected fishes are similar to those of Borg (1960) and Anderson and Conray (1969). Rare occurrence of this disease particularly during summer months reveals that this bacterium belongs to virulence category which has also been confirmed by experimental infection trials.

For going results of experimental infection trails reveal that there has been less mortality in fish but infected fish showed marked symptoms of myxobacteriosis. The fish exhibited normal behaviour in feeding and movement. These findings suggest that ambiant condition of the experiments are not supposed to be a significant stress on the challanged fish. It has also been noticed that most of the haematological investigations of disease fish and their variations in different blood parameters are helpful only when the diseases are diagnosed on the basis of symptoms these observations have accordance with Banerjee *et al.* (2002). For practical diagnostic utilization, the different blood parameters appear to be meaningful to compare the fundamental characteristics of different blood parameters of symptomatic fish to that of control one and has confirmity with the earliar studies in *Catla catla* (Banerjee, *et al.*, 2002), *Cirrhinus mrigala* (Prasad & Quireshi, 1995) and *C. batrachus* (Prasad and Verma, 2003).

The decrease in the number of neutrophils might be due to the phagocytic action of the bacteria which disfavour the formation of neutrophils. Lymphocytes and basophils result due to pathological conditions made by *Myxobacteria.* Similar results have also been reported in EUS infected *Mystus cavasisu* (Mastan and Qureshi, 2001) and EUS infected *Clarias batrachus* (Kumari and Akela, 2002). The lowering of Hb concentration, PC V (Packed cell volume) and RBCs have also been reported in naturally/experimentally infected fish by Vibriosis (Aderson and Conray, 1970 and Harbell *et al.*, 1979) or by other bacterial infections (Iwama *et al.* 1986 and Prasad and Verma, 2003).

The forgoing results also suggest that there are a progressive changes in the TEC. These changes are associated with rise in average Lymphocyte, Monocytes, Eosinophils and Basophils. The increase in these parameters depends on the intensity of Myxobacterial infection. It has also been found that there were rare occurrence of basophils in healthy fish. But this finding has decimilarities with the findings of Mastan and Qureshi (2001) who reported marked decline in lymphocytes percentage in BUS infected *Mystus cavasius.*

The blood value of two fresh water fishes *C. batrachus* and *H.fossilis* show significant changes when infected with *Myxobacterium.*

ACKNOWLEDGEMENTS

The authors are thankful to ICAR, New Delhi for financial assestance. Thanks are also due to Prof. T. A. Qureshi and Prof. K. B. Singh for their constructive criticism and healthy suggestions rendered in the preparation of this manuscript.

REFERENCES

Anacker, R. L. and Ordal, E. J. (1959). Studies on the Myxobacterium *Chondrococcus columnaris.* I. Serological typing. J. Bact. 78:25-32.

Anderson, J. I. W. and conroy, D. A. (1970). Vibrio disease in marine fishes. In Symposium on Disease of Fishes and Shellfishes (S. F. Sniezsko, ed.) Am. Fish. Soc. Spec. Publ. 5 : 266-272.

Banerjee, S. K., Patra, B.C., Bandyopadhyay, P. and Tiwary, A. (2002). Changes of the blood parameters in an Indian Major Carp, *Catla catla* Ham. due to Myxozoan Parasitic infection. J. Aqua. Biol. 17 (1): 79-84.

Benjamin, M. M. (1985). Outline of Veterinary Clinical pathology. 3rd ed. Kalyani Publisers, New Delhi. 5 - 316 P.

Borg, A. F. (1960). Studies *onMyxobacteria* associated with diseases is salmonid fishes. J. Wild. Dis. 8 : 1-85.

Conroy, D. (1972). Studies on the haematology of the Atlantic salmon *(Salmo solar* L.). Symposium Zoological Society of London. 30 : 101-127.

Dalmo, R.A., Ingebrigtsen, K. and Bogwald, J. (1997). Non-specific defence mechanism in fish, with particular reference to the reticuloendothelia system (RES). J. Fish. Dis. 20:241-243.

Harbell, S. C., Hodgins, H.O. and Schiewe, M. H. (1979). Studies on the pathogensesis of vibriosis in coho salmon *Oncorhynchus kisutch* (Walbaum). J. Fish. Dis. 2 : 391-404.

Hsu, T. C, Shotts, E.B. and Waltman, W. D. (1983). A selective medium for the isolation of yellow pigmented bacteria associated with fish disease. News letter for the *Flavobaterium-Cytophaga* group no. 3 : 29-30.

Iwama, G. K. ; Greev, G. L. and Randall, D. J. (1986). Changes in selected haematological parameters in juvenile chinnok salman subjected to a bacterial challenges and toxicant. J. Fish. Biol. 28 : 563-573.

Kumari, N. and Akela, B. P (2002). Effect of Epizootic Ulcerative Syndrome on haematology of *Clarias batrachus* (Linn). Indian, J. Environ, and Ecoplan6(l): 113-116

Mastan, S. A. and Qureshi (2001). Histo-pathological and Haematolololical investigations on EUS affected fish, *Mystus cavasius* (HAM). Ecol. Env. & Cons. 7(3) : 241-244.

Prasad, Y., and Qureshi, T. A. (1995). Incidences of columnaris disease in *Clarias batrachus.* Aqua. 2000 A. D. 275-282.

Prasad, Y. and Verma, V. (2003). Haematological investigations on *Clarias batrachus* subjected to Flavobacterial infection. Bion. 23 (2): 43-49.

Talib, V. H. and Khurana, S. K. (1995). Haematology for students. New Delhi, CBS Pub., 415 p.

Yamasita, H. (1967). Haematological study of a species of rockfish. II. Changes in moist content of blood, specific gravity, serum protein, haematocrit value and urea nitrogen level of serum in the specimens affected by ulcers. Bull. Jap. Soc. Scient. Fish. 33 : 995-1001.

CHAPTER 5

ENVIRONMENTAL APPRAISAL OF DEPLETING FISH DIVERSITY OF KASHMIR VALLEY

M. A. Khan

Division of Environmental Sciences,

S.K. University of Ag. Sci. Tech. Care: GPO. Box. 726

Srinagar-190 001, Kashmir, India

ABSTRACT

Since Heckel's work in 1838, the ichthyofauna of hilly region of Kashmir has been fairly well researched. The valley, with its pristine water systems, harboured rich and diverse fish-fauna, and accounts for more than 80% of total fish production of the State. In view of taxonomic uncertainities, the exact/authentic census of fish species is far from complete; though Kashmir claims as many as 42 species in the check-list of S. Nath, 1986.

There has been drastic decline in the fish diversity of Kashmir region. The introduction of exotic fast-growing species (e.g. common carp) has posed threat to indigenous Schizothoracine fish. The grass carp *(Ctenopharyngodon idelld)* and silver carp *(Hypophthalmichthys molitrix)* introduced in some Kashmir fish-farms by J&K Fisheries Department merits proper environmental impact assessment. Anthropogenic stresses, over the years, have also remarkably affected the fish biodiversity. Habitat destruction, indiscriminate exploitation for commercial gains, deleterious effect of mechanical weed harvesters, use of pesticides in the catchment, introduction of exotic fish species, unabated eutrophication of waterbodies and environmental pollution are the major causative factors of depletion of fish diversity in the region.

The paper examines the various anthropogenic pressures endangering the Kashmir ichthyofauna and calls for measures aimed at wise-use concept towards the conservation of fishery resources of the valley.

INTRODUCTION

Kashmir Himalayan Valley is blessed with unparalleled water resources sustaining myriad of fish fauna. Maintaining this ichthyodiversity gains paramount importance in view of mounting anthropogenic perturbations in their natural environment. The general factors influencing fish distribution and diversity include, *inter alia,* consideration of evolution and biogeography, the tolerance range of the species to environmental factors and predatory action and inter-specific competition. Throughout the world, evidences are emerging of dwindling diversity of ichthyofauna. Warren and Burr (1994) report that about 33% of the North American native freshwater fishes are endangered. Dehadrai & Poniah (1997) mention that 79 fishes of India are threatened. Kashmir water ecosystems are no exception to this global phenomenon. The present communication attempts to focus on major factors responsible for affecting the fish diversity of Kashmir.

KASHMIR ICHTHYOFAUNA AND DEPLETION TRENDS

Status of Fish Diversity in Kashmir

Kashmir valley holds some unique clean, deep and large mountain lakes as well as the flatland lakes and wetlands-, majority of them home to diverse species of fish which accounts for more than 80% of the total fish production of the State. A number of stresses, mostly man-made, detrimental to fish species in their once ideal environments have been observed to operate in aquatic ecosystems. The principal factors identified for the decline of fishes (Dehadrai & Poniah, 1997) include physical habitat loss, degradation or alteration, and chemical pollution or alteration. Palearctic fishes belonging to Schizothoracinae, Botinae and Nemachelinae are well represented in the ichthyofauna of Kashmir valley. Exotic trout and endemic Cyprinids are found to be predominant groups in the mountain lakes (Zutshi & Vass, 1970). A checklist of fish of Jammu and Kashmir State prepared by Nath (1986) includes as many as 42 species of fishes for Kashmir. Yousuf (2000) recorded 37 number of fish species. However, Khan (2000a) while describing aquatic biodiversity of Kashmir documented 40 fish species. Over the past few decade there have been various factors, acting in unison affecting the fish fauna in the valley.

Impact of Exotic Fish on the Local Ichthyofauna

Introduction of the exotic species for commercial gains has resulted in the loss of diversity. The introduction, in 1956, of the exotic common carp, *Cyprinus carpio* L. caused a sharp decline in the population and almost exterminated the schizothoracine fishes in Kashmir valley (Dehadria & Poniah, 1997; Zutshi & Gopal, 2000). Three varieties of *Cyprinus carpio* L. viz., scale carp, mirror carp and leather carp introduced in the valley led to the predominance of the exotic species over the indigenous fish population (Foteder & Qadri, 1974). The major reasons identified for such drastic shifts are:

- Food competition due to more or less identical food spectra. Studies (Yousuf, 2000) indicate that the similarity in food items varies from 31.3% between *Schizothorax esocinus* and *Cyprinus carpio specular is* to 51.7% between *S. niger* and *S. curvifrons.*

- Higher fecundity of exotic carp.

- Low fecundity of the Schizotnoracine fish causes low production number of young ones compared to the common carp.

- Spawning migration of Schizothoracine (except *S. niger)* from low reaches to high altitude streams.

- Better fertilization and growth rate

The introduction of grass carp *(Ctenopharyngodon idelld)* and silver carp *(Hypophthalmichthys molitrix)* in some Kashmir waters (e.g. Trigam sar) by the J&K Fisheries Department needs proper ecological monitoring and their impact assessment on indigenous fish. Studies by Sandhu *et al.* (1994) also indicate that *Cyprinus carpio* and *Hypophthalmichthyc molitrix* have alarmingly reduced the Mahseer *(Tor putitord)* fishery in Govindsagar reserviour. According to Yousuf (2000) exotic species have also been found to feed on the fry and fingerling stages of the local fish, especially Schizothoracine. For instance *Salmo trutta* L. and *S. gairdnerii* L. are true carnivores and consume significant quantities of fish.

Stress Impact of Catchment

Reclamation of water-bodies and their catchment for construction of buildings and agricultural purposes has destroyed the natural habitat of fishes. The use of pesticides, in cultivable land in and around the catchment, has added harmful chemicals to water bodies.

Many of these chemical pesticides are non-biodegradable and get accumulated at various trophic levels of food chain. Consumption of plankton by the fish adds some of the toxic substances into their system, which can be detrimental for the very survival of the fish population itself.

Due to the increase in human pressure, many of the natural water-bodies have either reduced in size or completely disappeared. Environmental deterioration of the water takes place by the addition of nutrients from anthropogenic perturbations in the catchment. Fishes are affected the most by the media in which they live. Because of deforestation and other detrimental human activities in the catchment areas of the flatland lakes like Dal, Anchar, Manasbal and Wular, most of the breeding grounds have undergone siltation. This has further aggravated the habitat conditions the consequence of which result in the decrease of fish population. The chemical and physical changes may induce alteration in the composition and population of plankton or the destruction of spawning grounds. The effluents may also affect through toxic derivatives in the sewage.

Impact of Eutrophication

Heavy accumulation of nutrients causes eutrophication. There are certain species of algae, noted for their rapid multiplication in waters, which impart specific colour (e.g. bluish-green, green, reddish brown) to water-bodies causing 'algal blooms' and this often is responsible for large scale fish-mortality. The native fish Schizothoracine in particular, find it difficult to cope up with the problem of eutrophication and quite often suffer heavy losses. The decay of organic matter with concomitant, consumption of the oxygen content of the water causing fish-asphyxiation. Little thought is being given to this problem in Kashmir despite frequent recurrence of algal blooms in some lake waters of Kashmir. The development of red-bloom with inherent potential of inducing toxicity in aquatic environments endangering fish is a sign of accelerated eutrophication (Khan, 2000b). The effluents coming from hotels and hospitals in the vicinity of water-bodies may also be harmful for the fish population.

Impact of Mechanical Weed Harvesters

Different lakes of Kashmir are characterized by the prolific growth of many macrophytes. *Ceratophyllum demersum, Myriphyllum spicatum, Hydrilla verticillata, Nelumbo nucifera* and many more form the major part of the aquatic vegetation in the valley. Thick growth of these have been causing considerable hinderance to swimming, diving and other aquatic-sport. To overcome the situation, selective deweeding has been employed in Dal lake over the past decade or so. As a consequence of this practice some undesirable changes have taken place disturbing the balance of the lake ecosystems (Zutshi & Tikku, 1990). Furthermore, studies carried out (Ticku & Zutshi, 1991) show that for every 100 kg of harvested weeds about 500 g of fish get tangled and thus lost from the ecosystem. Estimates show that about 7-8 tonnes of fish i.e. 33% of annual production is removed each year because of deweeding alone.

One of the indirect effects of large scale non-selective deweeding is temporary reduction of plant cover and removal of the bottom sand and gravel from the beds which otherwise may be useful for feeding and breeding of some indigenous fish species like *Schizothorax*. Hence the fish loss may be due to the loss of habitat, of which aquatic plants form an important component. Weed harvesting exposes both invertebrates and small fishes to increased risk of predation and as such have deleterious effect.

CONCLUSION

Apart from introduction of exotics, over-exploitation and stresses due to water pollution or anthropogenic pressures along the catchment have caused adverse impact on the fish diversity. Today the scarcity of the protein-rich food for the burgeoning population is haunting the mankind and urgent need is felt to conserve the ichthyofauna of Kashmir valley. Despite a number of causes identified for the perturbing decline in fish diversity, it has not been possible to undo the damage caused by each factor. Thus, holistic approach integrating the concept of sustainable development and conservation measures would go a long way in arresting the depletion of fish diversity of Kashmir.

REFERENCES

Dehadrai, P.V. & Poniah, A.G. (1997). Conserving India's fish biodiversity. *Int. J. Ecol. Env. Sci.* 23: 315-326.

Foteder, D.N. & Qadri. M.Y. (1974). Fish and fisheries and the impact of *Cyprimis carpio* L. on endemic fish. *J. Sci. Univ. Kash.* 2(1-2): 79-90.

Khan, M.A. (2000a). Status report on aquatic biodiversity in Kashmir Himalayan Valley. In: Proc. Second Natl. Workshop on Environ. Statistics. Hyderabad 6-7 April. pp. 96-103. Minst. Statis. Progrm. Impl. GOI, New Delhi.

Khan, M.A. (2000b) Anthropogenic eutrophication and red tide outbreak in lacustrine systems of the Kashmir Himalaya. *Acta hydrochem. hydrobiol* (Weinheim) 28: 95-101.

Nath, S. (1986). A checklist of fishes of Jammu and Kashmir state (India) with remarks on the Ichthyogeogrpahy of the state. *J. Zool. Soc. Ind.* 38: 83-98.

Sandhu, G.S., Tandon, K.K. & Johal, M.S.(1994)- Growth studies of an endangered fishes, *Torpottitora* (Hamilton), from Govindsagar (Himachal Pradesh) India. In: Threatened fishes of India.(eds.) D.V. Dehadrai, P. Das, : & S.R. Verma pp. 137-147. Natcon Publ. 4. Soc. Nature Conserv. Muzzafarnagar, India.

Warren, M.L. & Burr, B.M. (1994). Status of coldwater species of United States. Overview of an imperiled fauna. *Fisheries* 19: 6-18.

Yousuf, A. R. (2000). Impact of man on fish diversity in Kashmir. In : Environment, Biodiversity and Conservation, (ed.) M. A. Khan. pp. 439-449. APH. Publ. Corp. New Delhi.

Zutshi, D.P. & Vass, K.K. (1970). High altitude lakes of Kashmir. *Ichthyologica* 10: 12-15.

Zutshi, D.P. & Ticku, A. (1990). Impact of mechanical deweeding on Dal lake ecosystem. *Hydrobiologia.* 200/ 201: 419-426.

Zutshi, D.P. & Ticku, A. (1991). Effect of weed harvesting on the fish population of Dal lake. *Int. J. Ecol. Environ. Sci,* 17: 217-224.

Zutshi, D. P. & Gopal, B. (2000). Himalayan lake ecosystem : current issues and threat perceptions. Verh. *Int. Ver. Limnol.* 27: 2167-2170.

CHAPTER 6

EX-SITU CONSERVATION OF CHOCOLATE MAHSEER (*NEOLISSOCHEILUS HEXAGONOLEPIS* MC CLELLAND) IN 'SHALLY LAKE' IN ARUNACHAL PRADESH

Boni Amin Laskar , Azen Pujen & *Diptendu Das

SRF, NRCCWF(ICAR) project. Fisheries Experimental Farm, IDULI-792110, Arunachal Pradesh.Email:aminboni@rediffmail.com

**DFDO, Govt. of Arunachal Pradesh, Lower Dibang Valley District.*

Investigation on the population structure and ratio of various size group of Neolissocheilus hexagonolepis or chocolate Mahseer in shally lake located in the lower Dihang valley district of Arunachal Pradesh, revealed that the once introduced (1994-95) fish is thriving well with the IMCs and other exotic carps and have suitably adapted in the lake environment (max. temp 34° and min. temp 10°c) for breeding and further propagation. The habitat parameters and CPHPM of chocolate mahseer has been discussed in the paper.

The importance of mahseer as a world famous game fish is well known. Among the seven different recorded mahseers; *Neolissocheilus hexagonolepis* (McClelland) or chocolate mahseer is one of the endemic fishes of North East India. However, in recent years the destruction of habitat and breeding grounds by human intervention, construction of Dams, heavy rate of soil erosion and land sliding has threatened the chocolate mahseer stock in the rivers of Arunachal Pradesh. With a noble aim to conserve the declining population; few hundreds of fries and fingerings of chocolate and golden mahseer, collected from different rivers were introduced in a lake called "Shally Lake" by the DFDO in the year 1994-95.

The shally lake is situated in the geographical co-ordinates between N Latitude 28° 10', E Longitude 95°25' and alt 395 masl. The lake has been established with water area of 2 (two) hect. during the year 1975-76 by the state fisheries department. The lake is endowed with various species of cyprinid fishes, of which the culturable varieties like catla, rohu, mrigal, silver carp, grass carp, common carp and calbasu are being regularly released in the lake.

The present investigation on the population structure and catch per hour per man (CPHPM) of *N. hexagonolepis* was carried out to ascertain whether its population in the lake is expanding, contracting or is stabilized. The ratio of various size groups was also analysed to know the current reproductive status of the fish.

The fish specimens have been collected through experimental fishing with the help of mosquito-clothed drag net. The specimens has been collected through five different sampling at an interval of two days. All the specimens were examined to know the maturity status and have been measured for total length, standard length and total weight. The collected specimen were acclimatized in a circular cemented tank with continuous water supply.

The catch per hour per man(CPHPM) of drag net operated by three fishermen for five hours (one hour each day) was computed to be 1.505 kg. Altogether 140 number of *N. hexagonolepis* of length group 67 to 285 mm have been collected. The habitat parameters of the lake was found to be optimum with maximum temperature of 34° c and minimum of 10°c. the maximum wetted depth and minimum wetted depth of the lake is 12m and 8m respectively; while the lake is provided with 40% of riparian vegetation cover. The substratum of the lake is enriched with sand and pebbles (90%) with occasional bedrocks and boulders. The lake water is clear enough with light transferency upto 2.5m.. The junction of the lake and the mighty feeding channel of perennial ground water source makes the habitat colonizable for chocolate mahseer.

The assemblage of different ages and sexes gives the population a definite structure. The fish specimen collected has been categorized on the basis of maturity status in to young (70-100mm), maturing (130-200mm) and ripe(260-400mm). The sampling contain large proportion of young fishes (43' 82%) followed by maturing(34'99%) and ripe(15'66%).

Since no record were available with the fisheries department about the accurate quantity of the fish introduced. Against this background, the present study is found to be highly imperative. The observation revealed that *N. hexagonolepis* is thriving well with the IMCs and exotic carps . The occurrence of fingerling size of *N. hexagonolepis* in the lake justifies that the semi-rheophilic fish have suitably adapted to the lacustrine lake environment since its introduction in 1994-95 and have established natural population in the lake. The results from the present study not only brings out the abundance but also throws light on the expanding population of chocolate mahseer in the lake. The survey infers that the said lake would be an ideal site to study on the breeding behaviour and other life history pattern of chocolate mahseer.

REFERENCES

Agarwal, HN (1994) Endangered sport fishes of Assam. In Dehadrai, PV; Das ,P and Verma S.R (eds.) Threatened Fishes of India, NATCON publ. No. 4:209-212.

Chaturvedi S.K (1976) Spawning biology of Mahseer *Tor tor* (Ham) J. Bom. Nat. Hist. Soc., 73 (l):63-73

Das, S.M and Pathani, S.S (1978). Studies on the biology of Kumaon Mahseer, Tor putitora India, J. Anim. Sc., 48 (6) :461-465.

Hora, S.L (1937-43). The game fishes of India part VIII to XVII, Mahseers of India, J. Bomb. Nat. Hist. Soc., (41-44), 272-285.

Jhingran, V.G and Sehgal, K.L (1978) cold water fisheries of India, Inland Fisheries Society of India, Barrackpore, pp-239.

Joshi, C.B (2002) Artificial seed production of golden mahseer in kumaon uplands. In vass, K.K. and Raina, H.S (eds.) Highland fisheries and aquatic resource management NRCCWF publ. 275-283.

Kar, D; Laskar, B.A ; Nath D; Mandal, M and Lal siamliana (2002). Tor progenius (Mc Celland) under threat in River Jantinga, Assam, Sci & Cult. 68 (7-8): 211.

Madan Mohan (2002). Nutrition and feed development for cold water fishes. In Vass, K.K and Raina HS (eds.) Highland fisheries and aquatic resource management NRCCWF publ. 248-305.

Das, S.M and Pathani, S.S (1978). Studies on the biology of Kumaon Mahseer *Tor putitora* Indian J. Anim. Sc., 48(6): 461-465.

Hora, S.L (1937-43). The game fishes of India part VIII to XVII, Mahseers of India, J. Bomb. Nat. Hist. Soc., (41-43): 272-285.

Jhingran, V.G and Sehgal, K.L (1978) cold water fisheries of India, Inland Fisheries Society of India, Barrackpore, pp-239.

Joshi, C.B. (2002). Artificial seed production of golden mahseer in Kumaon uplands. In vass, K.K. and Raina, H.S (eds.) Highland fisheries and aquatic resource management NRCCWF publ. 275-283.

Kar, D; Laskar, B.A ; Nath D; Mandal, M and Lal stanliana (2002). *Tor progenius* (Mc.Celland) under threat in River Jantinga, Assam. Sci.& Cult. 68 (7-8): 211.

Sehgal, Kamal (2002). Nutrition and feed development for cold water fishes. In Vass, K.K and Raina H.S (eds.) Highland fisheries and aquatic resource management NRCCWF publ. 248-305.

INTRODUCTION

The increased use of chemical fertilizers ...
self sufficiency in food grains production ...
Every human born on this earth is now ...
dangerous chemicals from the moment of conception ...

The constant removal of aquatic weed by ...
pond weeds on the productivity while the aquaculture ...
of the system. The ecologists would ...
undisturbed either by extraneous inter...
interest of aquaculturist are best served by ...

A lot of work has been done on interactions ...
weeds (Chandel & Mehta 1996, Kanchan & Jay...
1989) but reports on aquatic weeds are very ...
Allelopathy is the impact of secondary metabol...
animals and/or microbes (recipient). The plants can...
and avoid elimination plants use defense mecha...
chemical factors. The toxins are generally located ...
vacuoles, leaf waxes, leaf resins, bud exudate ...
primary and secondary metabolic products although ...
also familiar to allelopathy scientists (Narwal & Tauro ...

Organic compounds or allelochemicals are rel...
algae and fern) affecting the growth and function of ...
exudation or by decomposition of litter. The allelo...
benzoic acids, cinnamic acids, phenolic acids, ...
thiopenes, juglone & cineoles (Rizvi & Rizvi 1...
vegetational patterning in the plant communities ...

CHAPTER 7

An Approach to Biocontrol for Fish Pond Management

Dr. J.P. Verma

Central Institute of Freshwater Aquaculture P.O.: Kausalyaganga Bhubaneswar-751002

INTRODUCTION

The increased use of chemical fertilizers in agriculture helped the country in achieving self sufficiency in food grains production. However, it has also polluted the environment. Every human born on this earth is now subjected to the constant danger of contact of dangerous chemicals from the moment of conception to death.

The constant removal of aquatic weed by manual or mechanical methods from the fish pond weaken the productivity while the aquaculturist would like to enrich the nutriant status of the system. The ecologists would perhaps, take up to keeps its ecological balance undisturbed either by extraneous interference or by internal imbalance of population. The interest of aquaculturist are best served by chemical or bio control of weeds.

A lot of work has been done on interaction studies between crops, crop plants and weeds (Chandel & Mehta 1990, Kanchan & Jayachandra 1980, Eyini *et al* 1989, Ayaz *et al* 1989) but reports on aquatic weeds are very few Molisch (1937) had coined the term Allelopathy for the impact of secondary metabolites of one plant (donor) on other plants, animals and/or microbes (recipient). The plants cannot change their location. Hence to survive and avoid elimination plants use defense mechanisms which are based on physical and chemical factors. The toxins are generally located at or near the leaf surface in trichinous, leaf vacuoles, leaf waxes leaf resins, bud exudate etc. Plant produce and store large amount of primary and secondary metabolic products although the latter are not only in bulk but are also familiar to allelopathy scientists (Narwal & Tauro 1996).

Organic compounds or allelochemicals are released from all group of plants (including algae and fern) affecting the growth and function of receiving species either by volatilization exudation or by decomposition of litter. The allelopathic compounds are largely derivatives of benzoic acids, cinnamic acids, phenolic acids, cumarins. hydroquinones, benzoquinones, thiopenes, juglone & cineoles (Rizvi & Rizvi 1992) The chemical compound alter the vegetational patterning in the plant communities Plant succession, seed preservation role in agro-ecosystem and plant defense are some of the other impacts of chemical compound. Deficiency of nutrients or nutrient stress, moisture deficit or temperature more or less than optimium, increase the production of allelochemicals / phytotoxins. Allelochemicals affect the neighbouring species by interfering with various plant process such as cell division, mineral uptake, respiration, photosynthesis, and chlorophyll synthesis. They may lead to succession. According to Rice (1984) allelochemicals released from plants are directly responsible for species composition of plant communities or genetic composition of associated plant population. The surrounding soil also contained similar toxins and are transmitted to individuals of neighboring plant species.

METHODS OF BIOCONTROL

The bio-control may be done by using plants, fungus, insects or animals.

Plants

Surface cover-Reduction of light intensity

Ahmad (1954), Philipose(1959) has recommended use of *Salvmia, Pistia, Lemna, Eichhornia crassipes.* The water surface of the pond are covered by these plants. It interfere in the penetration of sunlight in the water and thus checks the growth of weed.

Use of fungus

Different species of fungus *Cercospora,* when inoculated in aquatic weed, it spreads and kills the leaves and ultimately weed is destroyed.(Table-7.1)

TABLE -7.1

Name of weed	Name of fungus
Eichhornia crassipes	*Cercospora rodmanii* *Cercospora piaropi*
Ipomea aquatica	*Cercospora ipomoeae* *C.limorensis*
Trapa bispinosa	*C. trapea bispinosae*
Jussiea	*Cercospora jussiaeae*
Nymphea lotus	*Cercospora nymphaecea*
Impatiens	*Cercospora fukushiana* *C. nojmani*
Commelina	*Cercospora commelinae*
Cyperus	*Cercospora Cyperi-rotundi*
Elcocharis	*Cercospora carias*

Bloom / phytoplankton

(i) Bloom prevents growth of weed due to high turbidity

(ii) Prowse (1962) recommended introduction of sewage for the control of weed.

(iii) It has been observed that presence of phytoplankton e.g. *Scenedesmus bijugatus, Chlorella pyrenoides* suppresses the growth of macrophytes e.g. *Eichhornia crassipes* (Sharma 1985).

Introduction of Ottelia alismoides

K.V. Krishnamurthy and R. Indira (1984) observed that *Eichhornia crassipes* exhibits positive antagonism and tries to drift away from Ottelia. Introduction of Ottelia to water bodies infested with *Eichhornia crassipes* causes anatomical and morphological changes in the latter leading to its destruction. The whole cycle of changes, degeneration and destruction takes place within three to four weeks. At the same time it was also proved that the *Ottelia* is harmless to other aquatic plants, animals and microbes. Besides this *Ottelia* could be removed easily.

Powder of Cuscuta sp.

The powder of Cuscuta sp. (convolvulaceae) when spread on *Eichhornia crassipes* it inhibits growth and plant die within ten days.

Toxicity of allelochemicals

Pandey D.K.(1994) found that *Eichhornia, Salvinia, Pistia* were relatively more tolerant to phytotoxicity of the allelochemical when compared with other floating and submersed weeds. Thelethal dose varies as shown in Table-7.2.

TABLE-7.2

Allelochemical	Lethal at (ppm)	Lethal to aquatic weeds
p. Hydroxybenzoic acid	50-100	*Eichhornia, Pistia, Salvinia, Azolla, Lemna, Spirodela, Hydrilla, Ceratophyllum, Najas*
Anisic acid	50	*Azolla, Spirodela, Lemmna,*
	100	*Hydrilla, Naja*
Salicylic acid	50	*Spirodela, Lemna, Hydrilla, Ceralophyllum, Najas*
	100	*Azolla,*
Coumeric acid	100	*Lemna, Hydrilla Ceralophyllum, Najas*

(Pandey. D.K. 1994)

Aqueous Leachetes

Effect of aqueous leachetes (3% w/v) of three macrophytes i.e. shoot and inflorescence of Polygonum glabrum wild (Polygonaceae) *Phragmites karka* (Retz) Trin Exstend (Poaceae) and shoot, root and rhizome of *Nymphaea stellata* Wild (Nympheaceae) were found toxic to *Eichhornia* crassipes.

Horsley 1977 observed that shoots are more toxic than roots by *P. karka* and *N. stellata* Fletcher Renney (1963) have reported that leaves contain higher concentration of toxic material than other parts. The leaves are soaked in water @ 3gm/litre of water for 24hrs at 28'-40°c. It is passed through cheese cloth, filtered and allowed to settle for 8 hrs and used.

Animals

Bird

Hickling (1962) used *Kakhi cambell* a duck. It feeds on *salvinia.*

Insects

(i) Weevil — *Neochetina eichhornae* and *N. bruchi* are used @ 3000 weevils/ha for *Eichhornia crassipes.*

Neochetina was introduced, in India during 1982 by IIHR, Bangalore, from Australia and USA. Neochetina belongs to the order Coleoptera, family-Cureulioidea. It is brown in colour, long, elbowed antena, with three pairs of leg, 4.5-5mm X 2.5 *mm* body. Antenna folds into special grooves on the prominent snout which has mouth parts at its end. Egg hatch within 7-14 days. Larva makes a tunnel towards the petiole and feeds on the tissue. The tissue get rotten, leaves with feeding scars turn yellowish brown and ultimately dry.

The weevils can be easily multiplied in out door tank and pools and they establish readily underfield conditions after inoculative release. These weevils multiply fast and with increase in larval population drying up of leaf lamina starts from tips followed by complete browning of leaves and collapse of affected plant. These weevils are capable of migration by flight to nearby tanks up to 30kms.

It has been observed at Bangalore (1983-86) that 22,000 adults of *N.eichhornie* and 9000 adults *of N.bruchi* is required for 1000 ha infested area. The success was 80-95% in six tanks covering 900ha of water surface within 36 months.

The weevil were released in Kuttanad district of Kerala (1984), Lok Tak Lake in Manipur, Mula Mutha rivers in Pune, Water bodies of Nasik, Mandya, Mysore. At all places the technology was found effective, cheap, long lasting and environmental friendly.

(ii) *Erastroides curvifancies* Hampson (Lepidoptera)

It eats on Pistia stratiotes.

The number of larva has been observed to be high during November to February at 22-26° c water temperature. The early larval stage is the active feeding phase. The late instar larvae pupate in the deep furrows of leaves and petioles after feeding for 7-9 days

It has been observed that 90 sq. metre surface area is cleared within a month. It consumes about 1.5kg. of weed during this period. Other insect are mentioned in Table-7.3.

TABLE-7.3

Insects	Weed
Sameodes albiguttalis (Moth)	*Eichhornia crassipes (Center 1982)*
Cassida sp (Beetle)	*Ipomea aquatica*
Paulinia acuminata (semi aquatic grasshopper)	Aquatic fern
Cytrobagous salvhiiae	*Salvinia molesta* (Introduced in Kerala in 1984, effective control within 12-16 months of release but poor in dispersal and hence it has to be collected and released at new sites (Room 1986)
Bactra verutana (shoot boring moth)	*Cypenrus rotundus* (Frick KE 1979)

***Kachuga tectum* (Turtle)**

Kachuga tectum prefer nocturnal feeding. It has been observed that intensity of feeding is more during might than day. It consumes @127 gm d^{-1} in cistern cement (CIFRI News Aug. 83 Vol.6 (3&4) 816 gm *Kachuga tectum* consumed the following . These were in cistern of cement.

Lemna trisulcate	166.6
Ipomea aquatica	150.0
Hydrilla verticillate	146.0
Ceratophyllum demersum	70.0
Eichhornia crassipes	104.00
Total	636.6 gm

6 turtle of avg. 18gm weight consumed 20.4% of their body weight while 16gm weight consumed 35.8% their body weight.

Fishes

(i) *Pangasius pangasius*

It controls helminthes causing human infection, insects and gastropods of the pond

(ii) *Hypophthalmichthys molitrix* (silvercarp)

It checks on the development of bloom, members of Cyanophyceae, Bacillariophyceae, and decayed macrophytes.

(iii) *Cyprinus carpio* (common carp)

Cyprinus carpio consume weeds and uproot aquatic plants while digging the soil bottom. Pruginin (1968) has reported that pond stocked with *Cyprinus* @ 800/ha (total wt. 5000gm) consumed *Ceratophyllum, Myriophyllum* He further observed that filamentous algae were consumed by using *Cyprinus carpio* weighing 500-600 each stocked @ 250-300/ha.

(iv) Tilapia

T. melanopleura and *T. mossambica* consume weed at 20° c and above temperature. 2470 to 4940/ha stocking density, removes *Chara, Nqja* and other soft vegetalion. *T. melanopleura* consume diatomes while *T. mossambica* consumes planklonic epiphytic algae.

(v) Ctenopharyngodon idella (grass carp)

2,46,000 nos. of grass carp have controlled 65 miles long Karakum canal in Russia within two years (Aliev 1963).

Feeding becomes intensive at temperature above 16° C, the fish consuming a wide range of aquatic weeds. According to Hickling (1966) large fishes are able to masticate the leaves of toughland plants and fibrous grass. It can consume weed upto 40-70% of its own weight. The fish can consume weed 286% of their body weight (Singh et al 1967). 100kg of weed produces one kg of grass carp. Grass carp of 200-250 mm size are ideal for stocking in a perennial weed infested pond since the smaller one may be prayed upon by the predatory fishes. A few observations are mentioned in Table-7.4 - 7.5.

Table- 7.4

Size (cm)	Weight (gm)	Rate No/ha	Weed consumed	Period (day)	Growth in Length (cm)	Growth in Weight (gm.)
26.5	215	125	Hydrilla	21	40.3	837
24.4	210	3751964	Hydrilla	32	37.5	770

(Alikunhi & Sukumaran 1964)

Table 7.5

Stocking denisity (No.)	Weight Kg/ha	Weed consumed	Time taken
300-375	78.8-173	Hydrilla	One month
125-150	175-225	Duckweed	18kg/fish/day

(Alikunhi & Sukumaran 1964)

Grass carp generally remove the following weeds :

27-42 mm size grass carp consumes:

Wolffia, Lemna, Spirodella

Fingerling juvenile adult: grass carp consumes :

Azolla, Salvinia, Hydrilla, Naja, Ceratophyllum, Ottelia, Vallisneria, Neehmandra, Utricularia, Trapa, Myriophyllum, Limnophila, Potamogeton, Nitella, Chara.

Grass carp does not take:

Pistia, Nymphrides, Nymphea, Nelumbo.

CONCLUSION

Use of natural chemicals can save the environment from pollution. Major challenges before the chemists and aquaculturist is to synthesize and develop herbicides or pesticides from plants or microorganism as these are more easily degradable. Many commercially successful herbicides are already being developed from natural compounds.

With more rapid development of biochemical and other fast acting bioagent, biocontrol, which is an environmentally compatible, should be given first preference in weed management programme.

The discovery of suitable bioagents is very expensive, involving visits to native land of the weed in search of its natural enemies, followed by their subsequent screening at home, but its operational cost is rather limited.

REFERENCES

Ahmad, N. 1954 On water hyacinth and its control in fish ponds. Tech. Pap. Indo Pacif. Fish coun. (6): 1-3

Aliev, D.S. (1963). The status and prospects of pisciculture of phytophagus fishes in Central Asia and Khazakstan. In problems rybokhozisistvennogo ispol. Zovannia rastitel noiadnykh ryb v vodoemakh sssr, Ashkabed Akademiia Nauk Turkum SSR (in Russia).

Alikunhi, K.H. and Sukumaran KK 1964 Preliminary observations on Chinese carps in India Proc. Indian Acad. Sc. 60 B (3): 171-88.

Ayaz Sajjida F, Hussai I. Illahi and Bong Seipkill 1989. Allelopathic potential of *Adhatoda vassica,* NEES, Koren, J. Botany 32:109-119.

Center, T.D. 1982. The water hyacinth weevils. Aquatics 4(2) 8.

Chandel, Y.S. and Mehta PK. 1990 Nematicidal properties of leaf extracts of wild sage *(Lantana camana)* Indian J. Agri. Sc. 60: 781-82.

Charudattan, R 1982, Biological control of weeds with plant pathogenes, John Wiley and sons

CIFRI news. *Kachuga tectum* (Turtle) August'83 vol. 6 (3 & 4).

Eyini M, Jayakumar M & Pannir selvam S 1989 Allelopathic effect of bamboo leaf extract on the seedlings of groundnut Trop Ecol. 30: 138-141.

Frietcher, R.A. & Renney, A.J 1963 A growth inhibitor found in *Centauraea* sp Can J.Plant Sc. 43:475-481.

Frick, K.E et al 1979. Comparative control of purple and yellow nut sedge with B*actra verutana* Weed Sc. 27(2): 178-183.

Gopal, B & Goel U 1993. Competition and allelopathy in aquatic plant communities Bot Review 59: 155-210.

Hickling, C.F. 1966. On the feeding process in the white Amur Ctenopharyngodon idella J. Zool. 148:408-19

Horsley, S.B. 1977. Allelopathic interferences among plants II Physiological modes of action. In proceedings of the IV North American Forest Biology Workshop / Ed. Wilcox H.E. & Hamer AF: 93-136

IIHR Report 1982, Bangalore.

Jimenez, M. M., Lopez, EG, Delgadila RH & Franco ER 2001. Importance, rearing, release and establishment *of Neochetina bruchi* (Coleoptera curculionoidae) for biological control of water hyacinth in Mexico J. Aqua. Pl. Manage. 39:140-148.

Julien, M.H. & Griffiths, MW. 1998. Biological control of weeds a world catalague of agents and their targert weeds. 4th ed. CABI, publishing UK.

Kanchan, S.D. & Jayachandra 1980. Allelopathic effects of *Parthenium hysterophorous* IV-Identification of inhibitors. Plant and soil 55:67-75.

Krishnamurthy, K.V. & R. Indira 1985 weed vs. weed science for villages NO.92 May 1985.

Molisch, H. 1937. Uber der Einlfluss eiver pflanze auf die andere allelopathic, Gustav Fischer, Jena: 106 pp

Narwal, S.S. & Tauro P. 1996 Allelopathy field observation and methodology. Scientific publishers. Jodhpur.

Pandey, D.K. 1994. Inhibition of *Salvinia (S. molesta* Mitchell) by *Parthenium* (P. hyster phorus L) Relative effect of leaf, flower, stem, root, residue on *Salvinia* and paddy. J. of chemicals Ecology 20:3123-3131.

Rice, EL. 1984 Allelopathy 2nd edition Academic Press New York: 421 pp

Rizvi, S J.H., Rizvi V. (Ed) 1992. Allelopathy basic and applied aspects. UK-chapman & Hall

Room, P.M. 1986. Biological control in solving the world *Salvinia molesta* problems Weed Abs. 36 (2) 1987 item 511.

Sandbank, E & Hepher, B 1978. The utilization of microalgae as feed for fish, microalgae for food and feed: Ergels: Limnol 11:108-120.

Sharma, K.P. 1985 Allelopathic influence of algae on the growth of *Eichhornia crassipes* (Mart) Solma. Aquatic Bot.22:71-78.

Singh, S.B. 1966 Observation on efficacy of grass carp in controlling and utilizing aquatic weeds in ponds in India. Proc. Ind Pacific Fish coun 12(2) 220.

CHAPTER 8

BIOACCUMULATION OF CHROMIUM TRIOXIDE AND ITS EFFECT ON BLOOD GLUCOSE, GLYCOGEN CONTENT AND LDH ACTIVITY IN THE FISH, *TILAPIA MOSSAMBICA*.

R.Thangam and A.A. Sivakumar

P.G. & Research Department of Zoology, Kongunadu Arts and Science College, Coimbatore - 641 029, Tamilnadu, India.

ABSTRACT

Tilapia mossambica was exposed to sublethal concentrations of Chromium trioxide, 2,4,6,8 ppm for a period of 24, 48 and 72 hrs and changes in the biochemical contents have been observed. The accumulation of Chromium in musle, liver and kidney was found to be increasing with the increase of exposure period and concentration.The toxicity of heavy metal caused the blood glucose level to decrease with the increase of Chromium trioxide concentration and decrease with the increase of exposure period.The glycogen content in muscle and liver was found to decrease with the increase of concentration and decrease with the increase of exposure period. Increase in the Lactate dehydrogenase activity in the muscle; liver, kidney was observed with the increase in concentration of the heavy metal, Chromium trioxide, and exposure period.

Keywords : *Toxicity, Chromium trioxide, glucose and glycogen levels, LDH activity, Tilapia mossambica*

INTRODUCTION

There are several cases of Chromium contamination in ground water. The more serious of these cases are due to disposal of electroplating wastes in ponds or settling basins. The toxicity of Chromium to aquatic life is strongly influenced by the form of Chromium and quality of water. The toxicity towards Chromium considerably varies between groups of organisms. Invertebrates and phytoplankton have been found to be more sensitive than fish. Among plants, algae are apparently sensitive to Chromium than higher plants. The toxicity to algae is measured in terms of reduction on growth or in terms of reduction in photosynthesis. Freshwater invertebrates have a wide range of tolerance to Chromium and in general, they have been found to be more sensitive than fish. Experimental studies have suggested that calcium chromate may be that specific carcinogenic agent (Enterline, 1974). However, some investigations have produced cancer in experimental animals with injections of either the trivalent or hexavalent form (Hueper and Payne, 1962).

MATERIAL AND METHODS

The fingerlings of *Tilapia mossambica* were collected from the Department of Fisheries, Malampuzha in Kerala. Ten healthy fishes were exposed to different concentrations of the chromium trioxide to calculate the LC_{50} value (Finney, 1978). 1/10th of 24hrs LC_{50} value was used for the sub-lethal studies, in which the fishes live for long periods without any visible changes. The fishes were exposed to 24, 48 and 72hours. Control and treated fishes were sacrificed at the end of the exposure period. Blood and tissue samples (Liver, Kidney, and Muscle) were collected. The blood sample was used to analyse the glucose level. Other tissue samples were used to analyse the chromium level, glycogen, and Lactate dehydrogenase activity. The tissues, weighing about 20mg, were homogenized in tris buffer and centrifuged at 2500 rpm for 15 minutes. The clear supernatant was used for the analysis of all the parameters.

RESULTS AND DISCUSSION

In the present study, accumulation of Chromium trioxide in the tissues of the fish, *Tilapia mossambica* was found to be increased, with the increase of concentration of heavy metal. Higher amount of Chromium was present in the liver (Table 8.1). Van Hoof and Van San (1981) revealed that the rudd *(Scardinium erythrophtalmus)* bioaccumulated Chromium in the gill tissue, followed by the liver. Lowest Chromium accumulation has been observed in the muscle. Similar observations have been made by earlier studies (Nishihara *et al.*, 1985; Lazos *et al*, 1989; Seymore, 1994; Coetzee, 1996; Wepener, 1997).

Table 8.1

Accumulation of Chromium trioxide in the muscle, liver and kidney of the fish, *Tilapia mossambica*, exposed for different concentrations and periods.

Exposure concentration (ppm)	Muscle			Liver			Kidney		
	24hrs	48hrs	72hrs	24hrs	48hrs	72hrs	24hrs	48hrs	72hrs
2	0.030	0.1015	0.1146	0.060	0.1844	0.2266	0.042	0.1148	0.1398
4	0.068	0.1521	0.1534	0.021	0.2248	0.3128	0.086	0.1700	0.4416
6	0.071	0.1725	0.2029	0.032	0.2805	0.3827	0.096	0.2988	0.6144
8	0.089	0.1817	0.2564	0.037	0.3845	0.4761	0.130	0.3292	0.7354

Values are expressed in percentage

Each value represents the mean of three replications

Low accumulation of metals in muscles of some clupeiod fishes was reported by Sivakumar *et al.* (1989).The Cadmium accumulation was higher in the order of liver and kidney as reported by Thiruvalluvan *et al.* (1997). Chromium accumulation is highest in the order of liver and kidney. High levels of Cadmium accumulation were reported for liver, kidney of the trout, *Salvelinus fontinalis* by Benoit *et al.* (1976).

Blood glucose level was found to be increasing with the increasing of concentration and decreasing with the increase of exposure period (Table 8.2). Schwarz and Mertz (1959) observed the glucose tolerance factor, which was shown later to contain Chromium, was deficient in animals with impaired glucose tolerance, and that supplemental Chromium improved glucose tolerance; Glucose uptake, glucose use for lipogenesis, glucose oxidation to carbondioxide and glycogenesis increase because of the addition of Chromium plus insulin to animal tissues (Anderson, 1987). Diwan *et al,* (1979), Usha and Ramamurthi (1989) reported the imbalance in hormone release during heavy metal toxicity which probably blocks the absorption of glucose in to the tissue with consequential rise in blood glucose levels. Nath and Kumar (1987) and Bhattacharya *et al.* (1987) reported the fall in blood glucose level in fishes due to metal exposure.

Table 8.2

Effect of Chromium trioxide on the glucose level in the blood of *Tilapia mossambica* on exposed for different concentrations and periods.

Exposure concentration (ppm)	Exposure periods		
	24hrs	48hrs	72 hrs
Control	109	98	85
2	111	101	96
4	125	112	105
6	138	127	110
8	144	138	127

Values are expressed in mg / dl.

Each value represents the mean of three replications

Present study indicates the decrease of glycogen content in liver, muscle and kidney with the increase of heavy metal concentration (Table 8.3). Similarly, Banerjee *et al.* (1978) observed the depletion of glycogen content in liver after four weeks of exposure. Sastry and Subhadra (1982) observed the reduction glycogen content in the liver of the catfish, *Heteropnuestes fossilis.* Vijayaram *et al.* (1989) studied the Cadmium induced changes in the decreased glycogen contents in the fish tissues. A similar degree of decrease in glycogen content of Cadmium treated fishes was also reported (Shaffi, 1978; Dubale and Shah 1981; Lowe-Jinde and Niimi, 1984). Many reasons could be attributed to the decrease in tissue glycogen. Shaffi (1978) reported that accumulation of Cadmium in the pancreatic islets, of fish would selectively damage the insulin producing cells which in turn have contributed for the observed glycogen.

Table 8.3

Chromium trioxide on the glycogen content in the muscle, liver of the fish. *Tilapia mossambica* exposed for 24 hrs, 48 hrs and 72 hrs.

Exposure	Muscle			Liver		
concentration (ppm)	24 Hrs	48 Hrs	72 Hrs	24 Hrs	48 Hrs	72 Hrs
Control	0.926	0.927	0.925	3.768	3.766	3.764
2	0.766	0.526	0.385	1.309	0.931	0.729
4	0.763	0.484	0.340	1.205	0.867	0.660
6	0.758	0.471	0.295	1.555	0.835	0.596
8	0.694	0.419	0.265	1.091	0.803	0.510

Values are expressed in mg/g

Each value represents the mean of three replications

Activity of the Lactate dehydrogenase enzyme was increased in the increase of concentrations of Chromium and increase of exposure period of the fish, in the present study (Table 8.4). Fish favour anaerobic metabolism to meet the energy demands under metal stress as indicated by increased Lactate dehydrogenase activity in muscle as studied by the Vineeta Shukla and Sastry (1998). Similar study showed the increased activity of LDH in the liver of *Gambusia affinis* when treated with nitrate (Archana Kumta and Gaikwad, 1998).

Table 8.4.

Effect of Chromium trioxide on the Lactate dehydrogenase activity in kidney of the fish, *Tilapia mossambica* exposed for 24 hrs, 48 hrs and 72 hrs.

Exposure	Muscle			Liver			Kidney		
concentration (ppm)	24hrs	48hrs	72hrs	24hrs	48hrs	72hrs	24hrs	48hrs	72hrs
Control	0.128	0.129	0.128	0.131	0.133	0.129	0.119	0.122	0.121
2	0.149	0.156	0.194	0.172	0.201	0.256	0.132	0.150	0.174
4	0.153	0.166	0.214	0.185	0.205	0.269	0.140	0.151	0.179
6	0.156	0.171	0.217	0.188	0.219	0.277	0.147	0.155	0.204
8	0.161	0.181	0.226	0.196	0.238	0.291	0.159	0.158	0.228

Values are expressed in μ moles of pyruvate liberated min/mg. protein.

Each value represents the mean of three replications

Lactate dehydrogenase, which is a cytoplasmic enzyme, showed marked elevation in activity. LDH is generally associated with cellular metabolic activity. It acts as a pivotal enzyme between the glycolytic pathway and the tricarboxylase acid cycle. Thus non-availability of oxygen inhibition of LDH and simultaneous elevation of LDH suggest a bias towards anaerobic glycolytic pathway. This clearly indicates slower mobilization of glycogen from the liver (Larson 1975; Larson and Haux, 1982).

Lactate is the end product of the glycolytic sequence under anaerobic condition. LDH isozymes have high affinity for pyruvate as the electron acceptor. This accounts for high levels of LDH after treatment to nitrite, causing accumulation of lactate under tissue hypoxia causing their stiffening, affecting transamination. Inanov *et al.* (1977) also observed increase in lactate in the presence of Sodium nitrite.

REFERENCES

Archana Kumta and Gaikwad, S.A. 1998. Effect of Nitrite oh succinate dehydrogenase (SDH) and lactose dehydrogenase (LDH) in fresh water fish *Cambusia affinis.* Poll.Res., 17(2): 177-179.

Anderson, R.A. 1987. Chromium in trace elements in human and animal nutrition vol. 1,5th Ed., W.Mertz, ed, New York. Academic.press Inc., p: 225-244.

Banerjee, S.K., Dastidar, G. Mukhopadhyay,P.K., and Dehadri, P.V. 1978. Toxicity of Cadmium: A comparative study in the airbreathing *Tilapia mossambica.* Ind. J.Exp. Biol., 1274-1277.

Beneoit, A.D., Leonard, E.N., Christensen, G.M. and Fiandt, J.T. 1976. Toxic effects of Cadmium on three generations of Brook trout, *Salvelinusfoatinalis.* Trans. Amer. Fish. Soci., 105 (4): 550-560.

Bhattacharya,T., Ray, A.K. and S. Bhattacharya. 1987. Blood glucose and hepatic glycogen interrelationship in *Channa punctatus (Block).* A parameter of nonlethal toxicity bioassay with industrial polluntants. Indian. J. Exp. Biol., 25: 539-541.

Coetzee, L. 1996 Bioaccumulation of metals in selected fish species and the effect of pH on Aluminium toxicity in a cichlid, *Oreochromis mossambicus.* M.Sc. Thesis, Rand Afr. Univ., South Africa.

Diwan, A.D. Hingorani and Chandra Sekhram Naidu, N. 1979. Levels of blood glucose and tissue glycogen in two live fishes exposed to industrial effluents. Bull.Environ.Contam.Toxicol., 21:269-272.

Dubale, M.S. and Shah, P. 1981. Biochemical alternations induced by Cadmium in the liver of *Channa punctatus.* Environ. Res., 26: 110-118.

Enterline, P.E. 1974. Respiratory cancer among Chromate workers. J. Occup. Med., 16:523-26.

Finney, D.J. 1978. Statistical methods in biological assay. 3rd edition. Griffin Press, London, pp. 508.

Fruton, J.S. and Simmonds A. 1965. General biochemistry. Asia Publ. House, Madras.

Hueper, W.C. and Payne, W. W.1962. Experimental studies in metal carcinogensis. Chromium, Nickel, Iron, Arsenic. Arch. Environ. Health., 5:445-562.

Ivanov, N., Nestrov, N., Profirov, Y., Toucheva, E. and Petkava, O. 1977. The effect of Sodium nitrite on the ATP, ADP, AMP, and Lactic acid concentration in the liver, skeletal, heart musculature of rats. Shivotonovud Nauki. 14: 87-91.

Larsson, A. 1975. Some biochemical effects of Cadmium of fish in Koeman, J.H. and Strik JJTWA (ed) Sublethal effects on toxic chemicals on aquatic animals *Elsevier Amsterdam,* pp 3-13.

Larsson, A. and Haux, C.1982. Altered carbohydrate metabolism in fish exposed to sublethal levels of Cadmium. J. Environ. Biol., 3:71-81.

Lazos, E.S., Aggelousis, G. and Alexakis, A. 1989. Metal and proximate composition of the edible portion of 11 freshwater fish species. I food Comp. Anal., 2:371-381.

Lowe-Jinde, L. and Niimi, A.J. 1984. Short term and long term effects of Cadmium ion glycogen reserves and liver size in rainbow trout *Salmo gairdneri* Richardson. Arch. Environ. Contam. Toxicol., 13: 759-764.

Nath, K. and Kumar, N.1987. Toxic impact of hexavalent Chromium on the blood pyruvate of *Colisa fasciatus.* Acta Hydrochi. Hydrobiol., 15: 531-534.

Nishihara,T., Shimamoo,T., Wen, K.C. and Kondo, M. 1985. Accumulation of Lead, Cadmium and Chromium in several organs and tissues of Carp. Eisei Kagaku., 31(2): 119-123.

Sastry, K.V. and Subhadra, K.M. 1982. Choronic effect of Cadmium on some aspects of carbohydrate metabolism in a freshwater catfish, *Heteropneustesfossilis.*, Wat.Air Soil Pollut, 20: 293-297.

Schwarz, K. and Mertz, W. 1959. Chromium(III) and the glucose tolerance factor. Arch. Biochem. Biophys., 85:292.

Seymore,.T. 1994. Bioaccumulaiton of metals *in Barbus marequensis* from the olifants River, Kruger National Park and letal levels of Manganese to juvenile *Oreachromis mossambicus.* M.Sc., Thesis, Rand Afr. Univ., South Africa.

Sivakumar, R., Sivakuniaran, K.P. and Ramaiyan, V. 1989. Heavy metals in some clupeid fishes from Parangipettai waters. National Symposium on Natural Resources and their Conservation. Parangipettai.

Shafi, S.A. 1978. Cadmium intoxication on tissue glycogen content in three freshwater telosts. Curr. Sci., 47: 668-870.

Thiruvalluvan, M, Nagendran, N. and Charles Manaharan, A. 1997. Bioaccumulation of Caqmium and Methyl parathion in *Cyprinus carpio var. communis* (Linn). Journal of Environment and Pollution, 4 (3): pp.221-224.

Usha, R. and Ramamurthi, R. 1989. Histologrcal alternations in the liver of teleost *Tilapia mossambica* in response to Cadmium toxicity. Ecotoxicol. Environ. Safety., 17(2): 221-226.

Vanhoof, F. and Vansan, M. 1981. Analysis of Copper, Zinc, Cadmium and Chromium in fish tissues. A tool for detecting metal caused fish kills. Chemosphere ,10 (10): 1127-1135.

Vijayram, K., Geraldine, P. Varadarajan, T.S. George John and Loganathan, P. 1989. Cadmium induced changes in the biochemistry of an airbreathing *fish Anabas testudinus.* J.Ecobiol., 1(4): 245-251.

Vineeta Shukla and Sastry, K.V.I 998. Acute toxic effects of Cadmium on various parameters in the freshwater fish. Proc. Acad. Environ Biol., 7(2): 155-160.

Wepener,V. 1997. Metal ecotoxicology of the Olifants River in the Kruger National Park and the effect thereof on fish haematology. Ph.D. Thesis, Rand Afr. Univ., South Africa.

CHAPTER 9

HEPATO-TOXICITY OF ORGANOPHOSPHOROUS COMPOUNDMALATHION ON THE PROTEIN METABOLISM IN *CIRRHINA MRIGALA* (HAM.)

Vishal Tiwari

Department of Zoology, R.B.S. College, Agra-282002, U.P., India.

ABSTRACT

Cirrhina mrigala (Ham.) were exposed to subletted concentration of malathion (50 E. C) (2 ppm) for 7,14 and 21 days. After each exposure period, certain biochemical parameters were studied. A decrease in total, structural and soluble proteins and an increase in free amino acids and protease activity levels in contrast to protein decrement were noticed in 7 and 14 days of exposure, but on 21 days of exposure all the values comes to normalcy. The restoration of different protein fraction to normalcy indicates that after 14 days of exposure there seems to exist an oscillatory phase in protein turnover towards a more synthetic phase leading to the establishment of recovery and adaption phenomena.

Key words: *Malathion-toxicity, protein metabotishm, Cirrhina mrigala (Ham.).*

INTRODUCTION

The indiscriminate use of pesticides in agricultural operations affects the aquatic environment to a great extent. Extensive use of organophosphorus (OP) compounds has resulted in a widespread distribution of these chemicals in the environment. Due to rapid biodegradability and less persistency in the environment, the OP compounds replaced the more persistent organocholorine compounds and are produced in greater quantities now days (Anees, 1975). Some of these pesticides enter the aquatic ecosystem and adversely affect the aquatic biota like fish (Holden, 1972). The literature pertaining to the "Pesticide-protein interaction" makes a significant contribution in the field of pesticide biochemistry.

The organo-chlorine group of pesticides is more toxic, affects most diverse group of organisms, and persists in the environment for longer periods. Some of these compounds undergo bio-concentration and bio-magnifications (Rudd, 1964 and Wilber, 1969), the residues of organophosphate compounds during their short period of exposure affect the physiology of fish (Nagarathmma, 1982). According to Koundinya and Ramamurthi (1982), organophosphates altered the physiological activities such as tissue respiration and metabolism in the fish. Sreenivasulu Reddy *et. al.* (1986) reported decrease in total carbohydrate and glycogen levels in freshwater crab, *Oziotelphusa senex senex* exposed to malathion.

Since proteins are involved in the architecture of the cell and also in the physiology of the cell, they occupy a key role in cell parameters in the liver tissue of the nutritionally valuable food fish, *Cirrhina mrigala* exposed to commercial grade malathion. Thus, the present study was an attempt to study the hepato-toxicity of malathion on the protein metabolism in *C. mrigala.*

MATERIALS AND METHODS

The fish *Cirrhina mrigala* weighing 10 ± gram and average length 8.5 cm were collected from fish farm in the close vicinity of Agra. They were acclimated to laboratory conditions for a period of 15 days in glass aquarium. Fish were maintained in dechlorinated tap water with the following physico-chemical characters; dissolved oxygen 7 to 7.5 mg/ lit., salinity 0.6 ppm, temperature 26 ± 2°C, pH 7.5 ± 0.1, total alkinity 25.4 mg/lit, total hardness 112 mg/lit A photoperiod of 16 L; 8 days was maintained throughout. Fish were well reared and fed with rice bran along with Hitachi ready-made food pellets.

Preliminary test were carried out to find out the median tolerance limit (LC_{50}) of fish to malathion (50 B.C.) for 96 hrs, by Probit analysis method (Finney 1971). The concentration at which 50% survival/mortality occurred, was taken as the median lethal concentration (LC_{50}) for 96 hrs, which was 10.06 ppm., 1/5th of the LC_{50} value (2 ppm) was taken for the sublethal studies according to Sprague (1971). The control and malathion exposed fish were aerated frequently to prevent a hypoxic condition of the medium. After 7,14 and 21 days of exposure the liver tissue was isolated and used for the estimation of total proteins (TP), structural proteins (STP) and soluble proteins (SOP) (Lowry *et al.*, 1951). Free amino acids and protease activity (Morre and Stein, 1954) in control and malathion exposed fish at different exposure fish at different exposure periods and the mean values were subjected to statistical treatment using the analysis of variance (Pillai and Sinha, 1968). If the difference between any two mean values exceeds the critical difference (CD) the value are considered as significant. The CD values are given in Table 9.1.

Table 9.1

Level of total, structural and soluble proteins, free amino acids (mg/g wet wt.), protease activity (mg of amino acids per mg ptotein per hour) and the ratios of SOP/TP, STP/TP, SOP/STP in the liver tissue of control and Malathion exposed fish.

Parameters	Control	Exposure Periods (days) (7)	(14)	(21)	CD
Total Proteins (TP)	135.5 ±2.76	104.27* ±1.94 (-23.24)	76.31* ±1.98 (-44.07)	105.27* ±2.33 (-22.54)	3.01
Structural Proteins (STP)	90.46 ±2.12	61.84* ±1.81 (-40.02)	46.99* ±2.37 (-51.89)	69.31* ±.68 (-23.38)	2.66
Soluble Proteins (SOP)	44.27 ±4.39	39.88 ±3.04 (-9.91)	28.70* ±1.30 (-35.17)	34.73* ±3.07 (-21.54)	3.73
Free Amino Acid	4.937 ±0.086	5.034 ±0.080 (+1.98)	5.259* ±0.172 (+6.25)	5.687* ±0.144 (+15.19)	0.189
Protease Activity	0.553 ±0.021	0.569 ±0.01 (+2.89)	0.595* ±0.019 (+6.32)	0.429* ±0.012 (-22.42)	0.027
SOP/TP	0.325	0.382 (+17.53)	0.376 (+15.69)	0.329 (+1.23)	
STP/TP	0.665	0.593 (-10.82)	0.615 (-7.51)	0.658 (-1.05)	
SOP/STP	0.484	0.644 (+31.69)	0.610 (+24.74)	0.50 (+24.51)	

Note

1. Each value is mean ± S.D. of six individual estimations.
2. Values in the parentheses are percent change over control.
3. Indicates statistically significant at 5% level.

RESULTS AND DISCUSSION

The changes in the liver protein biochemistry of the fish exposed to malathion are presented in the Table 9.1. The total, structural and soluble proteins decreased, whereas free amino acids and protease activity levels were found to increase in fish exposed to 7 and 14 days. But in fish treated with malathion for 21 days all the values reached nearer to normalcy.

The decrement in the total, structural and soluble proteins suggest the existence of a high protein synthesis. The decrease in protein levels may be due to their degradation. The degradation products may in turn be fed into a tricarboxylic-acid cycle through the aminotransferase system to cope up with the high energy demands augmented during malathion stress (Malla Reddy, 1987; Bashamohideen, 1988). To have a more precise understanding of the variation among structural, soluble and total proteins, the ratios namely SOP/STP were calculated. The STP/TP values were correspondingly high over SOP/TP and SOP/STP values. These ratios clearly indicate that recycling of soluble proteins was considerably higher compared with structural proteins. However, the net protein budget was balanced at the total proteins level. Hypoprotenemia was observed in the selected tissues of fish exposed to organophosphorous pesticides by many investigators, thus supporting the findings of the present study (Ramalingam, 1885; Dev Prakash Raju, 2000).

The elevation office amino acids has a functional relevance so as to meet energy demands and is also involved in osmorgulation (Mutiny *et* al., 1984). The increase in protease activity under stress condition clearly suggests that malathion induces in protease activity which leads to the formation of higher 'free amino acid content causing hepato-toxicity. The liver is found to be affected more than any other tissue because it is the metabolic centre for detoxification and also the host in absorbing greater organophosphorous pesticide residues. The restoration of the protein level to normalcy indicates that after 14 days of exposure there seems to exist an oscillatory phase on a protein turnover towards a more synthetic phase or a less degradative phase leading to the establishment of recovery and adaptive phenomena.

REFERENCES

Anees, M. A., 1975. Acute toxicity of 4-organophosphorous insecticides to a freshwater teleost, *Channa punctatus*. Pak. J. Zool., 7:135-141.

Bhatt, S. L.; Bhatt, U. G. and Neelakantan, B. 1996. Acute toxic effect of the pesticide malathion on the larval and post-larval stages of the banana shrimp, *Penaeus merguiensis* (De man). Polln. Res. 15:281-283.

Bashamohideen, M. D. and Malla Reddy, P. 1987. Changes in brain proteins profiles of freshwater fish, *Cyprinus carpio* under malathion stress. Zeit-Ang. Zool., 74:393-397.

Devaprakash, Raju. 2000. Fenvalerate induced changes in the protein metabolism in the freshwater fish, *Tiapia mossambica*. Ph.D. Thesis. S. K. University, Ananntpur, Andhra Pradesh, India.

Finney, D. T. 1971. Probit Analysis. 3rd Edition. Cambrdge Univ. Press, London.

Holden, A. V. 1972. The effect of pesticides on life in freshwater. Proc. Royal. Soc. Lond., 180:383-394.

Koundinya, P. R. and Ramamurthi, R. 1982. Effect of organophosphorous pesticide Sumithion (Fenitrothian) an alkaline phosphatase activity of freshwater teleost, *Sarotherodon mossambicus*. Current Science. 15: 503-505.

Lowry, O. H.; Rosenbrough, N. J.; Fair, A. L. and Randall, R. J. 1951. Protein estimation by pholin phenol method. J. Biol. Chem. 193:265-275.

Malla Reddy, P. 1987. Modulation in protein profiles of white muscles of *Cyprinus carpio* under malathion stress. Biomed. 7:12-24.

Moore, S. and Stein, W. H. 1954. A modified ninhydrin reagent for the photometric determination of amino acids and treated compounds. J. Biol. Chem. 211:907-913.

Murthy, K. S., Kasi Reddy, B., Swami, K. S. and Chetty, C. S. 1984. Changes in respiration and ionic content of tissues of mussel exposure to methyl parathion toxicity. Toxicol Lett. 21:281-291.

Nagarathamma, R. 1982, Effect of organophosphorus pesticides on the physiology of freshwater teleost, *Cyprinus carpio,* Ph.D.). Thesis, S. V. University, Tirupati.

Pillai, K. K. and Sinha, G S. 1968. In statistical methods for biological workers, Agra, India, Ramprasad and Sons Pub.

Rudd, R. L. 1964. Pesticides and the living landscape (Wisconsin Press). 320-321.

Sreenivasvlu Reddy, P.; Bhagyalakshmi, A. and Ramamurthi, R. 1986. Carbohydrate metabolism in tissues of freshwater Crab, *Oziotelphusa senex senex* exposed to methyl parathion. Bull. Environ. Cantamn. Toxicol. 36:204-210.

Venkataramana, G. V.; Anandhi, U.; Shenbagavalli and Murthy, P. S. 2001. Histological changes in the cardiac muscle of the freshwater gobiid fish, *Glossogobius giuris* (Ham.) in response to malathion treatment. Poll. Res. 20: 407-411.

Wilber, C. C. 1969. The biological aspects of water pollution. Charles, C. Thomas Publications: 104-105.

Three Dimensional Analysis of Lentic Fish Biology

Manoja Kumar Mahapatra*

Environmental Biology unit, School of Life Sciences, Sambalpur University, Jyotivihar, 768019

ABSTRACT

Fishery science is a super science. Fishery survey has an influential space on the path of sustainable development. The three dimensional analysis including, ecobiology, economic viability, sustainable development, frequency, abundance, density, limnology, biophysical location and diversity indices of fishes have been correlated with each other. The lentic fishes are important than the lotic fishes as the 63% of the rural belly dependent on them. The fishery richness monitored the ecobiology of the region and attains a high human resource development index through economic viability and financial freedom.

Key words : Sustainable development, ecobiology, limnology, economic viability, diversity.

INTRODUCTION

Fishery Science is a super science. It has all branch of modern biology. Fishery science enjoys a very special consideration and place in human civilization from times immemorial. To study the fishery resources properly we have to think globally but act locally. Fishery survey furnished data on the spot information and needs sincere thinking, careful planning, hard work and dedication to the subject. It should have following objec-tives.

(i) Where does the present fishery stand ?

(ii) How much of the natural resources remain un exploited ?

(iii) What are the obstacles coming in the way of sustainability ?

(iv) What about the public participation and awareness ?

(v) What are the remedy for sustainable development ?

*Present address, Department of Zoology, Khariar College, Raj Khariar - 766107, Orissa, India.

The ever growing global food problem needs a momentus reply from fishery science. As fish has high protein value, the possibilites of fishery explosion should be explored at every level to give a sustainable reply to balance food problem. According to a FAO statisties countries may be classified on the basis of their per capita consumption of fish as follows :

(i) Very high i.e. 20 K.g. or more

(ii) High i.e. 10-20 K.g.

(iii) Medium i.e. 5-10 K.g.

(iv) Low i.e. less than 5 K.g.

India is placed in the last catagory i.e. in the class of low consumption countries. Keeping the growing realization of the importance offish as an alternate source of food, the Government of India scream for 'Blue revolution' after the successful Green revolution. India has enormour potential resources teeming with fish. The lentic (cultured) fishery resource is more important than the lotic (captured) fishery resource as 63% Indians (rural people) dependant on inland water harvesting structures (IWHS). The sustainable development of agroaquaculture is the successful management of resources for agroaquaculture production to satisfy changing human needs while maintainging or enhancing the quality of the environment and conserving natural resources (Mahapatra *et al* 2003). The biophysical, ecobiological and socioeconomic enviornment of India differs from region to region. Very little scientific information is available on the ecobiological attributes and production dynamics of the IWHS including traditional water harvesting struc-tures (TWHS) of India. It requires a teachnically appropriate, economically vialbe and socially acceptable technology for fishery producation (Mahapatra et al 2003, Mahapatra, 2004). Keeping all these factors in mind, the present study three dimensional analysis of lentic fish biology is conducted in Orissa.

MATERIALS & METHODS

The random samples were collected from different ecobiological, biophysical and geographical regions of Orissa. It comprises different IWHS excluding the natural water harvesting structures (NWHS) such as reservoirs, dam, lake etc. The limnological analysis were standardised from APHA (1995), Saxena (1990) and Ghosh *et al.* (1983). Fish samples were captured by different crafts and gears with the help of supporting villagers. The diversity indices were calculated by using Simpson (1949) and Shannon-weiner (1949) methods.

RESULTS AND DISCUSSION

Fishery culture in India is existed since 300 B.C. In India West Bengal, Bihar, Orissa and Andhra Pradesh are credited with age old practice in this trade. Till date the enchancement of fishery is not satisfactory. The per capita consumption is too less than the global average. The path of development and expansion of fisheries lay a number of obstinate dogmass and custums. The tropical ecology prevading the resources and the tropical climate of the land, however are a serious disadvantage to Indian fisheries. The sparse fishery expansion may be due to the following reasons:

(i) Ecobiological location

(ii) Climatic impacts and constraints

(iii) Financial inviablity

(iv) Poor limnological status

(v) Energy and inputs

(vi) Nutrient availability

(vii) Genetic resources

(viii) Fish mortality

(ix) Policy and management

(x) Public awareness

To improve human resource development (HRD) index, sustainable use of local resource is needed. It is easy to integrate all these needs and resources skillfully and behavioural adjustments on the other thereby demanding a process of new learning and unlearning leading to a real ecodevelopment, (Mahapatra, 2002, Mishra & Dash, 1981).

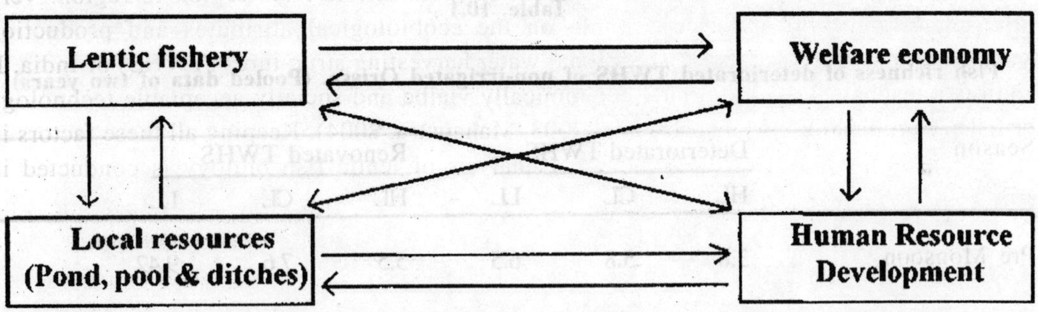

The three dimensional analysis of fishes comprises the analysis of fishes from following dimensions.

(i) Ecobiology, Economic and sustainability.

(ii) Frequency, Abundance and Density.

(iii) Limnology, Biophysic and Diversity.

Ecobiology, Economic and Sustainability

Every organism of the universe may have different habitat but a niche specific life on ecobiological point of view. Like wise lentic fishes have a niche specific survivility. They may be surface living, bottom dwelling or living in the transtional area. The ecobiology of the fishes (may be of same species) vary according to the nature of their IWHS habitat. According to Mahapatra, 2004 the IWHS is divided into three types.

(i) Natural water harvesting structures (NWHS) such as lake, dam, reservoir etc.

(ii) Traditional water harvesting structure (TWHS) such as *kata, muda, sagar, ghaghra, bundh*

(iii) Artificial water harvesting structure (AWHS) such as *chahla,* paddy fields etc.

The survivility of fishes, their energy requirement (both climatic and nutrient status) and their niche specific life differs from one ecosystems to other. The major carp fishes were mostly occurred is NWHS, a few TWHS such as *kata, sagar* and *bundh* and absent is AWHS (Mahapatra etal. 2003).

The *sagar* has more species diversity while the *chahla* has less (Mahapatra & Mishra 2003). As the present study is not included the NWHS, the ecobiologically observed fishery diversity is as follows :

Paddy fields < Chahla < Muda < Ghaghra < Kata < Bundh < Sagar

The species diversity is more in perennial (wet) ponds. The pollution of the environment and human exposure affect the ecobiological status of fishes. The deteriorated pond has less diversity as compared to the renovated ponds as depicted in Table 10.1 of Mahapatra & Mishra 2004.

Table. 10.1

Fish richness of deteriorated TWHS of non-irrigated Orissa. (Pooled data of two years)

Season	Deteriorated TWHS			Renovated TWHS		
	HL	CL	LL	HL	CL	LL
Pre Monsoon	3.8	5.8	6.5	5.5	7.6	9.42
Monsoon	5.41	6.32	9.21	8.1	9.28	10.6
Post Monsoon	4.37	5.98	7.46	7.3	10.6	9.89

The fish has a high economic value. The welfare economy of a region should be baked by local resources. In Orissa the welfare economy should be backed by fish ecobiology. The ecologial aspects especially the IWHS should be managed well for rice and fish production. (Mahapatra, 2004). Sinhababu *et.al.* (1996) observed that rice-fish system can annually produce 16% of food crops, 0.6% fish and prawns besides 5.2% of fodder (straw). Das *et.al.* (1996) reported that the net returns as well as cost : benefit ratio were maximum in rice-fish culture. Ghosh *et.al.* (2003) also reported the economic viability of rice and fish culture. Thus an attempt should be made in Orissa to promote rice and fish culture.

Due to the less efficient of the dual culture (rice & fish), the Orissa in general and Western Orissa (Koshal) in particular suffers from malnutrition and starvation death.

The questionnary of Mahapatra & Mishra 2004, speaks the following dimensions of peoples (Table 10.2) having fish diversity (renovated TWHS) and less fish diversity (deteriorated TWHS).

Table-10.2
Available data from questionnary

Sl. No	Subject	Deteriorated TWHS	Renovated TWHS
1	Crop diversity	39%	47%
2	Draught loss	53%	21%
3	Crop productivity	48%	64%
4	Farmers indebtness	27%	16%
5	Pisciculture	13%	35%
6	Horticulture	2%	4%
7	Agro Forestry	9%	7%
8	Malnutrition	59%	28%
9	Starvation	0.59%	0.29%

Integration of ecobiological objectives into fishing policies and management with involvement of fishermen and other relevant stake holders is needed to improve economic viability (Hopkins, 1999). The economic viability and financial freedom of a community may lead to the high HRD index of the region. It can lead to sustainable development of the community.

Sustainable development in an integral process. The key concept of sustainable development is the sustanable use of natural resources which allows long term economic growth and enhancement of production (Mahapatra & Mishra, 2004). Fisheries is an integrated component of the ecosystem. Principles guiding sustainable fisheries include utilization of the ecosystem in a manner consistent with productivity, biodiversity, germplasm, conservation, application of the precautionary approach to the management of living resources and the mod-ern technology to enhance the growth and diversity of fishery resources. In the study area not only the traditional Indian major carps but also the non conventional vulnerable fish species need proper conservation, cultivation and attention. The proper limnological condition for the better sustainable development of the fisheries is depicted in Table 10.3.

TABLE 10.3

The best Physico-Chemical condition for higher fishery diversity, (after Mahapatra, 2004)

Sl. No.	Parameters	Overall ranges	Diversity Low	Medium	High
1.	Water temperature	12-31 c	18	19-22	>31
2.	PH	6.5-9.0	<6	6-8.5	>9
3.	Dissolve Oxygen (mg/l)	4-14	<4	6-10	>12
4.	Specific Conductivity	76-474	-	up to 200	>200
5.	Free carbon-dioxide (mg/l)	3-7	<3	4-5	>7
6.	Total Alkalinity (mg/l)	76-474	<40	40-90	>90
7.	Total Nitrogen (mg./l)	3-7	<0.5	1	>1.5
8.	Fluoride (mg/l)	40-230	<0.05	0.05	>0.5
9.	Available phosporous	0.05-1.5	<0.2	0.5-0.7	>1
10.	Organic Carbon	0.05-1	<0.5	0.5-1.5	>3
11.	Available Nitrogen	20-75	<20	20-40	>70
12.	Inorganic phosphorous	2-10	<0.5	1-5	8

In the field of fishery conservation and management, besides the economic viable species (India major carps) the sustainability locally available species (non carp) is needed to enhance the alternate economic viable sources and to fill the rural belly which inturns increases the HRD index of the community. The less available species *Notopteres chitala* and *Chela untranchi* is source of fish food throughout Orissa in rainy season. These fishes along with *Wallago attu* occurrence is maximal in monsoon. It is also available in post monsoon period in paddy fields of rice agro aquaculture. In rural Orissa farmers needs sustainable development of Paddy fields as life support systems through traditional technology of irrigation (by TWHS) for an assurance of availability of rice (Carbohydrate), fish (protein), Vegetables (Vitamins & Minerals).

Frequency, Abundance and Density

Frequency indicates the nos. of sample unit on which a given species occurred and this express the distribution of various species in a community ecosystem.

The Indian major carp species have more frequency in monsoon and pre monsoon, while the non carp species have more frequency in the post monsoon period. This may be due to absent of proper climatic and limnological condition. (Table 10.4)

Table 10.4

Frequency, Abundance & Density of Major Carp(MC) & Non Carp (NC) Species (pooled data of 2001 & 2002)

Season	Frequency		Abundance		Density	
	MC	NC	MC	NC	MC	NC
Pre Monsoon	15.47	12.21	4.31	7.59	1.9	1.87
Monsoon	30.43	28.29	6.65	8.34	2.43	3.59
Post Monsoon	27.34	38.45	5.87	8.79	2.37	4.58

Table 10.5

Lentic fishery diversity observed in rural ponds of Western Orissa (pooled data of two years).

Sl.No.	Name of the Species	Sl.No.	Name of the Species
1.	*Notopteres notopteres*	16.	*Clarius batrachus*
2.	*Notopteres chitala*	17.	*Mystus vittatus*
3.	*Gadusia chapra*	18.	*Mystus aor*
4.	*Catla catla*	19.	*Wallago attu*
5.	*Labeo rohita*	20.	*Anabas testudieneus*
6.	*Labeo bata*	21.	*Pseudotropis authroriosides*
7.	*Labeo calabasu*	22.	*Channa punctus*
8.	*Cirrihinus mrigald*	23.	*Mastacembalus aculeate*
9.	*Cirrilunus reba*	24.	*Liza carsula*
10.	*Chela untrachi*	25.	*Channa striatus*
11.	*Punitus palehellus*	26.	*Rita rita*
12.	*Panitus sarana*	27.	*Tilapia mossambicus*
13.	*Danio spinousus*	28.	*Cyprinus carpio*
14.	*Labeo gonius*	29.	*Ctenopharyngodon idella*
15.	*Rasbora daniconius*		

Table 10.6

Frequency, abundance, density and diversity indices of inland fishes present in study sites of Orissa (Pooled data of two years).

Sl.No.	Frequency	Abundance	Density	Simpson's index	Shannon -Weiner's index
1.	23.07	8.33	1.92	0.0008	0.0012
2.	07.69	4.02	0.53	0.0001	0.0002
3.	15.38	13.01	2.51	0.0009	0.0017
4.	23.07	9.00	2.17	0.0008	0.0016
5.	53.83	9.42	5.07	0.0059	0.0063
6.	46.14	7.21	3.53	0.0018	0.0029
7.	30.76	5.5	1.69	0.00001	0.00003
8.	38.45	6.2	2.57	0.0015	0.0019
9.	23.32	7.9	1.09	0.0007	0.0011
10.	61.52	5.12	3.15	0.0025	0.0031
11.	38.45	15.2	4.87	0.0098	0.0081
12.	23.09	9.13	2.12	0.001	0.0015
13.	30.06	5.02	1.05	0.0005	0.0024
14.	46.14	5.58	2.87	0.0019	0.0018
15.	38.45	4.61	1.76	0.0009	0.0027
16.	15.38	10	1.5	0.0011	0.0019
17.	55.83	4.19	2.23	0.0015	0.0017
18.	15.38	3.5	0.53	0.00006	0.00018
19.	38.45	4.0	1.5	0.005	0.0024
20.	40.42	5.9	2.5	0.0012	0.0019
21.	17.69	8.51	0.61	0.00008	0.0031
22.	31.23	7.8	3.3	0.0012	0.0017
23.	38.75	4.28	2.12	0.000001	0.000003
24.	30.76	9.07	1.69	0.0019	0.0025
25.	30.76	9.07	1.69	0.0019	0.0025
26.	31.59	6.12	2.87	0.000001	0.000005
27.	23.04	7.57	2.03	0.0682	0.0358
28.	48.52	16.09	5.92	0.0026	0.0032
29.	46.14	8.03	4.81	0.0025	0.0032

Species abundance or species richness is the total nos. of species present. Abundance of various species has been expressed as percentage cover in the sample areas.

Density determine the size of population. It was expressed as nos. of individuals per meter square.

The abundance of non carp species is high than major carp species while the abundance of the individual species *Cyprinus carpio* is high than other major carp. The abundance of *Mystus aor* is low than other species. The species *Notopteres chitala* has less density in comparision to other.

The frequency, abundance and density of the fishes vary depending upon their different ecobiological, climatic and limnological conditions.

Limnology, Biophysici and Diversity

Limnology of water harvesting structure determines the community biodiversity. The limnology is influenced by the biophysical location (i.e. Foot hill, Central table and Riverbank) of the study sites.

The limnological characterics of all the sampling sites are depicted in Table 10.7.

Table 10.7

Pooled data of two years (i.e. 2001 and 2002) showing X ± SE of limnological Parameters from different biophysical sites of Orissa.

Param Foot hill (FH) ponds Central label (CT) ponds River bank (RB) ponds

	Pre-M	Monsoon	Post-M	Pre-M	Monsoon	Post-M	Premonsoon	Monsoon	Post-monsoon
Temp. of water °C	38.57±3.18	34.56±3.89	30.28±2.33	37.59+3.33	36.07±2.18	29.72±1.57	35.43±3.43	33.21±5.01	30.82±2.89
pH of Water	9.82±1.28	7.59±0.23	8.57±0.31	9.21±0.59	8.29+0.84	7.82+0.81	8.12+0.29	7.19+0.12	6.89±0.23
pH of soil	6.52±1.28	6.47±0.92	6.21±0.12	7.36±0.17	7.21±1.27	7.52±0.59	7.85±1.21	7.39±0.49	7.29+.1.19
Free Carbon dioxide ppm	138±18.3	98.57±10.21	9321±8.17	113.57±21.02	91.37±8.21	85.16±13.21	92.55±11.48	89.21±8.21	75.02±10.12
Total Alkalinity ppm	3.04x10² ±1.08	2.59x10² ±1.51	1.24x10² ±0.32	4.8x10² ±0.14	3.2x10² +0.51	1.79x10² ±0.38	5.2x10² ±1.02	3.92x10² ±1.21	2.19x10² ±0.89
Dissolved oxygen ppm	8.35±1.87	33.12±8.92	23.58+8.52	23.58±5.57	26.98±5.28	21.54±2.54	16.57±1.24	38.57+7.23	36.24+6.91
C/N of soil mg/100g of soil	0.6362± 0.0151	0.521 ±0.0121	0.672 ±0.0126	0.672 ±0.0240	0.582 ±0.0191	0.749 ±0.0182	0.257 ±0.0098	0.238 ±0.0021	0.254 ±0.0026

The adverse species diversity of the region may be due to the harsh limnological condition. For the better diversity of the fish the IWHS should fulfill the prescribed limnological condition by Mahapatra 2004. The biophysical location of the water bodies (FH,CT & RB) also affect to the limnology leads to alternate of the species diversity. The river bank area has more species diversity while foot hill area has less species diver-sity. For the sustanability of fishery science, ecobiological specific fish species cultivation i.e. hill stream fishes in foot hill, non carp species in central table and carp species in the river bank water bodies should be encouraged. It will not only enhance the economic viability but also add to germplasm conservation. All the three dimension of the fishes correlated with each others. Thus proper care should be taken about the three dimension at the time offish management, conservation, cultivation, production, preservation and the modern branches of fishery science.

Table 10.8

Fish Diversity in different Water Bodies of Orissa. (After Mahapatra, 2004)

Fish Diversity		Simpson Index			Shannon-Weiner index		
		Pre-M	Monsoon	Post-M	Pre-M	Monsoon	Post-M
Major-carp	FH	–	–	0.0024	–	–	0.0031
	CT	–	0.0012	0.0026	–	0.0018	0.0039
	RB	–	0.0059	0.0067	–	0.0065	0.0082
	FH	–	0.0011		–	0.0016	-
Non-Carp	CT	–	0.0012	0.0033	–	0.0018	0.0041
	RB	0.0025	0.0019	0.0026	0.0032	0.0024	0.0039

REFERENCES

APHA (1995) :- Examination of water and waste water. American public health association, Newyork.

Das, R.N ; Laha, W.K ; and Mazumdar, B. (1996) - *Studies on technology on paddy-cum-fish and prawn farming in low lying costal zone of West Bengal.* J. Indian Soc. Costal Agric. Res. in : 229-232.

Ghosh, A.B; Bajaj S.C; Hasan R. and Singh D (1983). Soil and water testing methods of IRRI.

Ghosh, M; Maiti; D and Nandi A.K (2003). *The economic viability of dual culture (rice + fish) in coastal saline zone of West*

Bengali, curr. sci. 3(2): 387-390. Hopkins. C (1999) - Sustainable Development Vol-31, N-3.

Mahapatra M.K. (2002) - *Impact of monsoon on growth of capitals in Hirakud command area* Proc UGC Seminar on the growth of capital in Hirakud command area, Problems & Possibilities held in Attabira College, Attabira. Oct. 2002 pp 112- 115.

Mahapatra M.K. & Mishra A (2003) -*Effects of limnological parameters on indigenous inland fishery diversity In western belt of Orissa* jr.current sciences Vol 3 No 2. pp 467-470

Mahapatra M.K., Mishra A & Mishra H.S. *(2003) -Comparative analysis of frequency abundant and density of available major carp fishes with non carp fishes of western Orissa. Indian* J. Env & Ecoplan Vol 7 No 3 (in press)

Mahapatra M.K. (2004) - Inland water harvesting structures and lentic fishing diversity in Orissa. Submitted to National Seminar on inland water resources and Environment University of Kerala Thiruvantapuram (accepted).

Mahapatra M.K. & Mishra A (2004) - Traditional water harvesting structures and sustainable development in non-irrigated areas of Orissa, India. Indian J. of Traditional knowledge (accepted). Mishra, A & Dash M.C. 1981). Education for survival and development. Proc. International Seminar on Environment Educations New Delhi Dec. 1981.

Saxena M.M. (1990) Environmental analysis. Agro biological publication New Delhi, India.

Sinhababu D.P. ; Jha ; K.P; Mathur, K.C and Ayappan S. (1996) Integrated farming system. Rice-fish for low land ecologies. Proc. Int. symposiun on Rainfed rice for sustainable food security. CRRI culture Sept. 23-25.

CHAPTER 11

A STUDY OF MATING AND BREEDING TACTICS OF *GAMBUSIA AFFINIS* (BAIRD & GIRARD 1853)

Vishal Tiwari, K.K Gaur and Rita Rani

Department of Zoology, R.B.S. College, Agra-282002.

ABSTRACT

The present investigation aimed to study the mating and breeding tactics of *Gambusia affinis* (Baird & Girard 1853). Rapid sexual maturity, prolific breeding habit, sex recognition, courtship and pair formation were some of the most important aspects of findings. Gonopodium as a secondary accessory sexual dimorphic feature in male along with gravid spot, ovoviviparity and superfaetation (producing number of broods from one mating) in female fish were promiscuous features of reproductive biology of *G. affinis*.

Key Words: ***G.affinis: mating-breeding tactics.***

INTRODUCTION

The fish of the Poeciliidae family are the best known of all aquarium fish due to their relative ease of keep and the many colourful cultivated varieties available. *G. affinis* is one of the most renowned, cosmopolitan fish in terms of its larvicidal nature throughout the world. It is of great biological, ornamental, educational, commercial value, and thereby having its peculiar importance.

Perhaps this has certainly triggered off many hobbyist, commercial breeders, exporters to systematically search for attractive species and varieties through artificial breeding techniques. A good knowledge on the biology, feeding behavior and ambient environmental condition of the fish species is a prerequisite for culture techniques. Keeping this in view the present study on *G. affinis* has been under taken to work out detailed reporductive behaviour. Some important contributions on the said field of studies are from Baylis (1981); Wourms (1981); Emlen. *et al*, (1997); Evans *et al* (1999); and that of Pilastro *et al* (1999).

MATERIALS AND METHODS

A set of 30 healthy adult male and female *of G.affinis* species of the same age group in the ratio 2:1 respectively were put in a glass aquarium of 24 X 12 X 12 inches filled with potable water. Netting, bagging, transporting and introducing fishes into an aquarium was done very carefully. Fishes that were healthy, having effortless swimming action and more active with clearly defined colour patterns on them, were chosen, eliminating diseased fishes and those with visible anatomical deformities in order to obtain the most vigorous fry.

The aquarium used was equipped with adequate illumination, meeting the requirements of the plants, filter, a thermostat, an aerator pump and the accessories. Another small sized aquarium of 12 X 6 X 6 inches was also prepared in the same manner possessing all the essential accessories as a breeding tank, to separate the females into it after the mating. Water quality parameters were monitored as they are considered to be the important factors influencing spawning and successful rearing of fry.

RESULT AND DISCUSSION

The males of *G. affinis* were noted comparatively smaller, more brighter, colourful and as a polygamous active members for mating preferences. The males were continuously in courtship display, always chasing after the females. The males were observed persistent drivers, so mating occurs frequently. Males sooner or later were send darting around a female, his fins spread and quivering. Acceptance of a male's advances is often signalled by head-shaking on the part of the female, followed by a dipping of the anterior part of her body (with a corresponding raising of the posterior half), which thus renders her genital aperture more accessible to the male once he adopts the correct mating position.

In order to fertilize the female, the anal fin of the male *G. affinis* developed into a structure known as "gonopodium", which inserts the packets of sperms into the female fish. It acts as the most significant external discriminatory feature between male and female and thereby, facilitating internal fertilization.

The males were noted to court the female by approaching her and bending his body into 'S' shape. This type of behaviour was repeated. As the males wings the gonopodium forward to make contact, the female will often fold her anal fin out of the way helping him even further. When mating, advancing from behind, male brought his gonopodium to a forward or sideways position due to its flexibility and insert it into the vent of the female.

After successful mating the females transferred to the breeding tank showed their chest fills out almost upto the head and her middle gets quite distended, as the days of delivery approaches. Females about to give birth were recognized by their round tight stomach, which in late pregnancy become cuboidal in shape. In the mature gravid female, slightly forward to the normal shaped anal fin inside her body appears a dark area known as the gravid spot. It was infact considered to be the womb. As the eggs incubate, the eyes of fry are sometimes visible through the thin walls of the gravid spot. (Fig. a, b, c and d)

About 4-8 weeks after mating, the females gives birth upto 10-30 youngs. At birth, the anal fin of the male *Gambusia* appears much like that of the female. As he grows, the fin becomes gradually modified so that in a resting stage, it appears narrow and somewhat longer than that of the female. At the age of 30-35 days the sexes can be distinguished by the development of larger body of the female and her gravid spot which males do not show. The fry were noticed gaining fast maturity and within 3-4 months they get ready for taking part in a successful mating. The highest state of evolution of parental care and breeding system is shown by viviparous teleosts (Baylis, 1981).

An interesting feature noticed in females that a fertile female have as many as half a dozen broods even if no males were present in the breeding tank. This behaviour may be attributed as a precautionary measure and to store the packets of sperms ejected by the male and uses them to fertilize a series of eggs and hatches as they ripen, following a single mating, females produces a number of fry at regular interval. This phenomenon is called superfaetation.

The wide range of tolerance to physico-chemical parameters of water (Table 11.1) and its adaptability and resilence along with superfaetation has made them suitable for aquarium life.

Table 11.1
Reproductive biology of *Gambusia affinis*

Sexual characteristics	Gestation period.	Size	Temp	pH	Hardness	Number of youngs delivered
Males		3-4 cm				
(i) Smaller in size						
(ii) Brighter, Colourful						
(iii) Active sexual partner						
(iv) Gonopodium						
(v) Attaining fast sexual maturity						
Females	4-6 weeks	5-7 cm	20-28°C	Slightly acidic 6-7	179ppm	10-30
(i) Comparatively larger						
(ii) Dull coloured						
(iii) Pot-bellied						
(iv) Gravid spot						
(v) Prolific breeders						
(vi) Ovoviviparity & Superfaetatioin						
(v) Attaining fast sexual maturity						

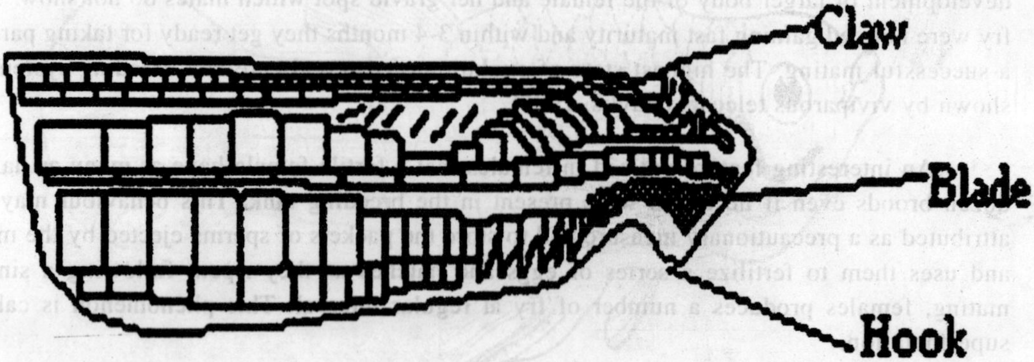

Fig. (a)

Abbreviations
Gonopodium of male *Gambusia affinis*.

Fig. (b)

Abbreviations
Gravid female fish of *Gambusia affinis*.

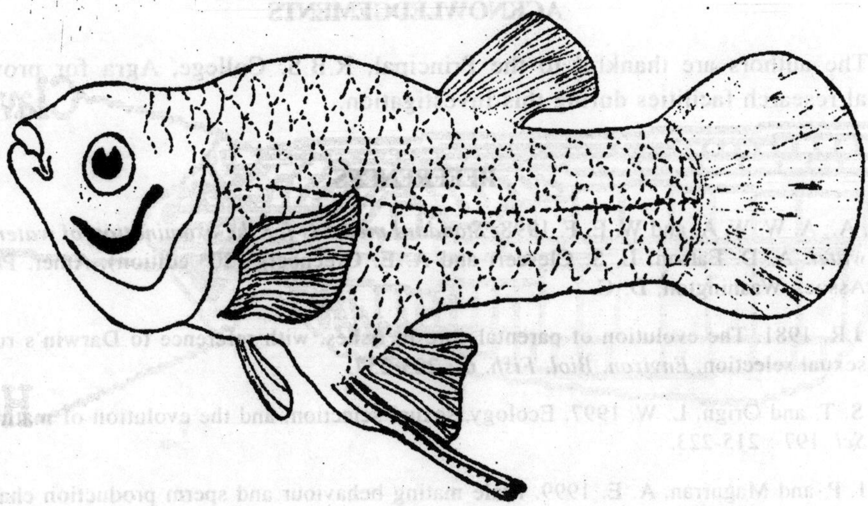

Fig. (c)

Abbreviations
Male with modified gonopodium

Fig. (d)

Abbreviations
Livebabies of *Gambusia affinis* (Mosquitofish).

ACKNOWLEDGEMENTS

The authors are thankful to the Principal, R.B.S. College, Agra for providing the essential research facilities during this investigation.

REFERENCES

A. P. H. A.; A. W. W. A. and W. E. F. 1998. *Standard methods for the examination of water and waste water.* A. D. Eatson, L. S. Clesceri and A. E. Greenberg (20th edition). Amer. Pub. Health. Assoc., Washington, D. C.

Baylis, J.R. 1981. The evolution of parental care in fishes, with reference to Darwin's rule of male sexual selection. *Environ. Biol. Fish. 6* : 223-257.

Emlen, S. T. and Orign, L. W. 1997. Ecology, sexual selection, and the evolution of mating systems. *Sci.* 197 : 215-223.

Evans, J. P. and Magurran, A. E. 1999. Male mating behaviour and sperm production characteristics under varying sperm competitions risk in guppies. *Anim. Behav.* 58 (5): 1001-1006.

Pilastro, A. and Bisazza, A. 1999. Insemination efficiency of two alternative male mating tactics in the guppy, *Poecilia reticulata. Proc. Royal. Soc. Biol. Sci. Series.* 266(1431): 1887-1891.

Wourms, J. P. 1981. Viviparity : the maternal fetal relationship in fishes. *Am. Zool.* 21 : 473-515.

CHAPTER 12

STUDIES ON QUALITATIVE AND QUANTITATIVE PLANKTON PRODUCTION USING DUCK DROPPINGS

S.K. Majhi, B.K. Mahapatra, K. Vinod and B.K. Mandal

Division of Fisheries, ICAR Research Complex for NEH Region, Barapani-793103, Meghalaya, India.

ABSTRACT

Two different doses of duck droppings i.e.0.25g/lt and 0.5g/lt were applied for plankton culture. On computation it has been found that, a dose of 0.5g/lt was ideal for both phytoplankton and zooplankton production. Among the phytoplankton, *Spinogyra* sp. and *Euglena* sp. and among the zooplankton, *Brachionus* sp. and *Lepadela* sp. were found dominating. The quantitative analysis revealed a phytoplankton population of 1229 nos./50 lit and a zooplankton population of 2225 nos./50 lit on the 30th day. On the 60th day, phytoplankton population was 2404 nos./50 lit and zooplankton population was 13810 nos./50 lit.

Keywords: *Duck excreta, Phytoplankton, Zooplankton*

INTRODUCTION

The rationale behind integrated farming is to minimize wastes from the various subsystems on the farm: wastes or by-products from each subsystem are used as inputs to other subsystems to improve the productivity and tower the cost of production of the outputs of the various subsystems (Edwards *et* al., 1986). Integrated livestock fish farming system is a proven environmentally sustainable and economically viable technology that encompasses rational utilization of available resources and livestock such as pigs, ducks, chicken, etc (NACA, 1989). Different forms of integrated livestock fish farming *viz.* pig-fish, poultry-fish, duck-fish etc. have been evolved and popularized in India (Sarma and Pattanaik, 1985; Das, M.K., 1990; Ayyappan *et* al., 1998; Borah *et* al., 1998; Ghosh *et* al., 2000 and Naskar et al., 2002). Recently attention has been paid to develop the live feed culture utilizing the animal excreta from poultry, pig, duck, rabbit etc. The live feed containing balanced amino acid is an indispensable requisite for larval rearing of the captive bred fishes, prawn, ornamental fishes etc. (Mahapatra and Datta, 1999; Mahapatra *et al.*, 1999 and Mahapatra *et* al., 2000). In North Eastern Region of India, livestock farming is a prime activity of majority of the rural households. One or more of the animal components like piggery, poultry and duckery are invariably found in all houses (Bujarbaruah *et. al.*, 1996). The social and economic conditions prevailing in NE region, integration of livestock may be the only source of fertilizer availability, at low cost, to make fish culture economically feasible and profitable.

Under this rural farming setup, fish farming would be more profitable if integrated with animal components and recycling animal excreta for the cause of plankton production and growth of the aquaculture industry. As the region has a very good scope for indoor aqua business especially for ornamental fish culture, the necessity of live feed culture technique is need of the hour. The present study was done to observe the qualitative as well as the quantitative plankton production using duck droppings.

MATERIALS AND METHODS

Six cement cisterns (C1-C6) were used for the plankton culture. C1 and C2 were used as control, C3 and C4 for treatment I (E1), C5 and C6 for treatment II (E2). Pond water was used in all the experimental cisterns. Manuring was not done in control while in treatment I (E1), duck droppings @ 0.25 g/lit and in treatment II (E2), duck droppings @ 0.50 g/lit were applied. A water volume of 90,30,000lt was maintained in all experimental cement cisterns. The study was carried out for a period of 60 days.

The plankton and water sampling were done on 30th and 60th day at 10:00 Hrs. from each treatment and control. Air temperature, Water temperature, pH, Alkalinity, Hardness, Dissolved Oxygen, Free Carbon-di-oxide, Chloride and Total Dissolved Solids (TDS) were determined quantitatively using *Standard Methods* (APHA, 1989). The plankton samples were collected by filtering 50 litres of water from each cement cistern using a standard plankton net. The live samples were brought to the laboratory for preliminary observations and then were fixed in 5% formalin and stored in glass bottles for detailed qualitative and quantitative estimation of plankton and the results have been presented based on the pooled data of the replication.

RESULTS AND DISCUSSION

The qualitative analysis of plankton and physico-chemical properties of the water of the experimental cisterns are presented in Table 12.1 and 12.2, respectively. It has been observed that the dominance of zooplankton population increased with that of higher dose of duck droppings. In E2, the dosage of manure was high and the zooplankton reached its maximum both during 30th and 60th day sampling, which is also supported by the findings of earlier worker (Tsirtsis and Karydis, 1998). The quantitative analysis of phytoplankton is presented in Table 12.3 and zooplankton in Table 12.4. It has been found that in control the net plankton production were 1569 and 1478 numbers, out of which phytoplankton population were 1407 and 1355 and zooplankton population was 162 and 123 on 30th and 60th days, respectively. In E1, the net plankton population was 1762 and 4639 numbers, out of which the phytoplankton population was 773 and 1170 and zooplankton population was 989 and 3469 on 30th and 60th days. Similarly in E2, the net plankton population was 3454 and 16214, out of which phytoplankton population was 1229 and 2404 and zooplankton population was 2225 and 13810. It has been found that on 30th day among phytoplankton the population of *Scenedesmus sp* was dominating in control and *Spirogyra sp* in E1 and E2. Among zooplankton *Keratella sp* was dominating in control and *Brachionus sp* in E1 and E2. Similarly on 60th day sampling, among phytoplankton *Scenedesmus sp* was found dominating in control and *Spirogyra sp* in E1 and E2. Among zooplankton, *Testudinella sp* in control and *Brachionus sp* in E1 and E2 were dominating. The quantitative variation of phytoplankton and zooplankton is presented in Figure 12.1 and 12.2.

This base line information on plankton production using duck droppings definitely helps in developing package of practice on live feed culture of plankton under controlled condition.

Table 12.1.

Qualitative plankton production in different experimental cisterns

Cement cistern	Phytoplankton	Zooplankton
Control	*Scenedesmus sp., Pediastrum sp., Ankistrodesmus sp., Ulothrix sp., Spirogyra sp., Staurastrum sp., Tetrastrum sp., Placus sp., Fragilaria sp.*	*Keratella sp., Arulla sp., Nauplius larvae, Testudinella sp., Polyarthra sp., Calanoid copepod.*
El	*Scenedesmus sp., Pediastrum sp., Ankistrodesmus sp., Ulothrix sp., Spirogyra sp., Staurastrum sp., Tetrastrum sp., Placus sp., Fragilaria sp. Cosmanium sp., Nitzschia sp.*	*Brachienus sp., Lepadella sp., Trichocera sp., Keratella sp., Arulla sp., Nauplius larvae, Testudinella sp., Polyarthra sp., Calanoid copepod.*
E2	*Scenedesmus sp., Pediastrum sp., Ankistrodesmus sp., Ulothrix sp., Spirogyra sp., Staurastrum sp., Tetrastrum sp., Placus sp., Fragilaria sp. Cosmanium sp., Nitzschia sp., Microsystis sp., Botryoccus sp., Ceratium sp., Navicula sp., Cryptomonas sp.*	*Brachionus sp., Lepadella sp., Trichocera sp., Keratella sp., Arulla sp., Nauplius larvae, Testudinella sp., Polyarthra sp., Calanoid copepod., Cyclopoid copepod., Aureopsis sp., Philedema sp., Alora sp., Arulla sp., Discotracha sp.*

Table 12.2

Physico-chemical parameters in different experimental cisterns

Water parameters	Experimental tanks	Air temperature (°C)	Water temperature (°C)	pH	Specific conductivity (µS/cm)	Dissolved oxygen (ppm)	Free carbondioxide (ppm)	Total alkalinity (ppm)	Total hardness (ppm)	Chloride (ppm)
On 30th day	Control	21.5±1.5	27.0±1.5	6.52	19.8	6.0	14.1	5.1	12.019	0.01086
	E1	21.5±1.5	27.0±1.5	6.53	20.1	6.0	14.1	5.0	12.012	0.01086
	E2	21.5±1.5	27.0±1.5	6.53	20.1	6.0	14.0	5.0	12.012	0.01065
On 60th day	Control	24.0±1.5	27.5±1.5	7.2	11.50	8.3	12.0	5.0	8.009	5.85
	E1	24.0±1.5	27.5±1.5	7.3	11.61	9.5	12.0	5.0	8.008	5.85
	E2	24.0±1.5	27.5±1.5	7.3	11.60	9.8	12.0	5.0	8.008	5.85

Table 12.3

Quantitative record of Zooplankton (no./50 lit) during the study period

Species	Quantitative record of Zooplankton on 30th day			Quantitative record of Zooplankton on 60th day		
	Control	E1	E2	Control	E1	E2
Dicranophorus sp	07	12	20	05	10	28
Euchlamus sp	10	00	00	00	08	10
Conochiloides sp	13	18	27	10	15	35
Colurella sp	12	10	16	09	05	20
Calanoid copepod	00	03	02	00	00	01
Cyclopoid copepod	01	00	01	00	00	02
Nauplius	02	00	00	05	00	00
Macnothrix sp	01	01	01	00	03	04
Diffilugia sp	00	00	01	00	00	01
Polyarthra sp	02	10	20	00	15	25
Testudinella sp	19	158	186	25	78	180
Lecane sp	00	03	00	00	00	01
Arulla sp	00	02	01	00	00	00
Cilates	00	00	00	00	00	01
Brachionus sp	12	527	1019	10	1992	7855
Alora sp	00	08	06	01	00	07
Philedema sp	00	02	00	02	00	00
Lepadella sp	05	220	830	07	1320	5525
Trichocera sp	00	00	01	00	00	05
Aureopsis sp	00	02	03	01	00	09
Keraiella sp	78	13	90	48	18	100
Discotracha sp	00	00	01	00	05	01
Total	162	989	2225	123	3469	13810

Table 12.4.

Quantitative record of phytoplankton (no./50 lit) during the study period

Species	Quantitative record of Zooplankton on 30th day			Quantitative record of Zooplankton on 60th day		
	Control	E1	E2	Control	E1	E2
Scenedesmus sp	1000	68	25	680	63	102
Ankistrodesmus sp	130	08	15	380	07	00
Pediastrum sp	200	90	43	173	45	212
Selanastrum sp	09	00	01	05	00	00
Ulothrix sp	10	00	01	04	00	00
Staurasitrum sp	00	00	00	00	00	03
Coelastrum sp	00	01	08	01	05	20
Tetrastrum sp	01	02	00	05	06	00
Placus sp	00	00	01	00	00	00
Fragillaria sp	00	00	05	00	01	15
Navicula sp	04	00	08	00	05	25
Cryptomonas sp	00	04	00	00	00	00
Dictyosphaerium sp	00	00	01	00	00	00
Euglena sp	01	53	140	05	41	280
Spirogyra sp	20	280	589	50	700	980
Microsystis sp	08	83	124	10	100	290
Botryoccus sp	04	100	82	01	57	225
Cosmanium sp	01	55	154	00	95	200
Ceratium sp	19	29	32	41	45	52
Total	1407	773	1229	1355	1170	2404

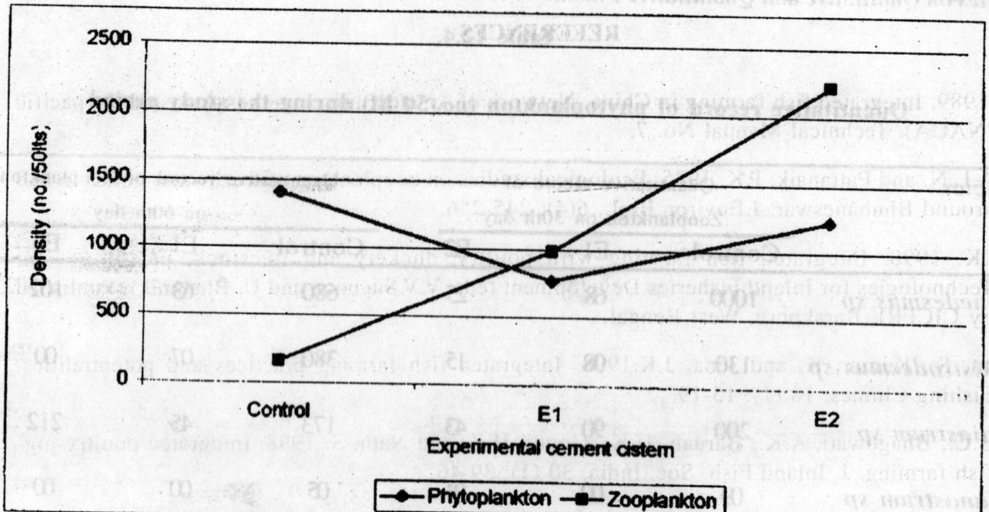

Fig. 12.1 Phytoplankton and Zooplankton on 30th day

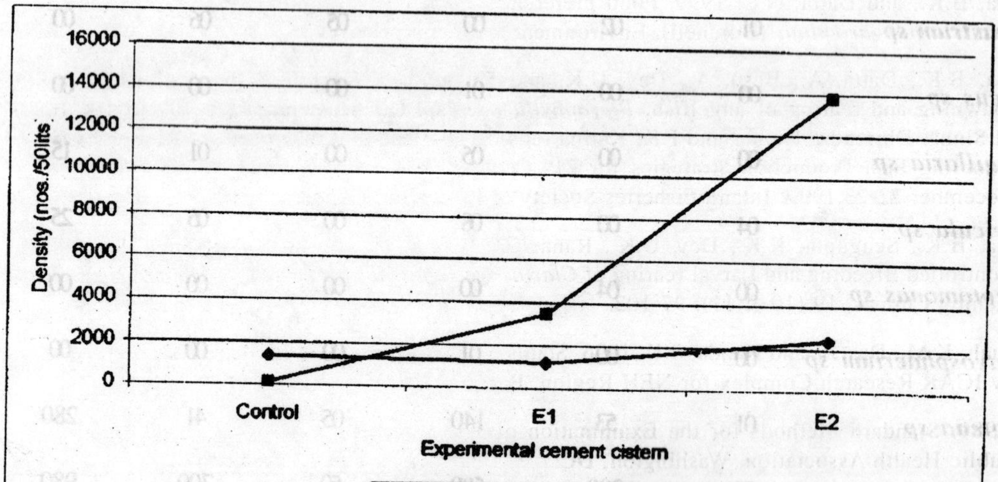

Fig. 12.2 Phytoplankton and Zooplankton on 30th day

ACKNOWLEDGEMENTS

The authors are thankful to the Director, ICAR Research Complex for NEH Region for providing necessary facilities to carry out the work and also to the authorities of NATP-CGP for financial assistance.

REFERENCES

NACA. 1989. Integrated fish farming in China. Network of Aquaculture centers in Asia and the pacific (NACA). Technical Manual No. 7.

Sarma, A.L.N. and Pattanaik, P.K. 1985. Ecological studies on zooplankton of freshwater ponds in and around Bhubaneswar. J.Environ.Btol., 6(4): 245-256.

Das, M.K. 1990. Integrated fish farming with poultry, duckery and livestock, pp. 59-68. In: Technologies for Inland Fisheries Development (eds. V.V.Sugunan and U. Bhaumik). Published by CICFRI, Barakpore, West Bengal.

Ayyappan, S., Kumar, K. and Jena, J.K.1998. Integrated fish farming practices and potentialities. Fishing Chimes. 18 (1): 15-19.

Borah, B.C., Bhagowati, A.K., Baruah, U.K., Gogoi, H.K. and Nath, S. 1998. Integrated poultry-pig-fish farming. J. Inland Fish. Soc. India, 30 (1): 39-46.

Ghosh, A.K., Mahapatra, B.K. and Baguti, T.K.. 2000. Integrated approach for enhancing fish production. Souvenir: State level Seminar-cum-Workshop on integrated approach for enhanced fish production. Feb. 24-25, Dept. of Fisheries. Govt. of Tripura.

Naskar, S., Mahapatra, B. K. and Das, A. 2002. Eco-friendly management of pig waste through integrated Pig-Fish farming - A case study. National Seminar on "Environmental Biology and Fish Biology", February 01-03, 2002, Visva-Bharati, Santiniketan : 64-65.

Mahapatra, B.K., and Datta, N.C. 1999. Food preference and growth performance of the exotic cat fish *Clarias gariepinus* (Burchell). Environment and Ecology, 17 (1): 207-210.

Mahapatra, B.K., Datta, A., Basu, A., Dey, U.K. and Sengupta, K.K. 1999. Observations on the spawning and rearing of angelfish, *Pterophyllum* sea/are (Lichenstein). Page 102 to 106 In: M.Sinha, Dhirendra Kumar and P.K. Katiha (eds), Eco-friendly Management of Resources for Doubling Fish Production-Strategies for 21st Century. Proceedings of the National Seminar, December 22-23,1999. Inland Fisheries Society of India, Barrackpore.

Mahapatra, B.K., Sengupta, K.K., Dey, U.K., Rana, G.C., Dutta, A., Basu, A. and Saha, S. 2000. Controlled Breeding and Larval rearing of *Clarias batmchus* (Linn.) for mass-scale propagation. Fishing Chimes, 19 (10 & 11): 97-102.

Btijarbaruah, K.M., Das, A. and Nanda, S.K. 1996. Status of animal husbandry in Meghalaya. Published by ICAR Research Complex for NEH Region. Barapani. Meghalaya.

APHA, 1989. Standard Methods for the Examination of Water and Wastewater, 17th edn. American Public Health Association, Washington, DC.

Tsirtsis, G. and Karydis, M. 1998. Evaluation of phytoplankton community indices for detecting eutrophic trends in the marine environment. Environmental Monitoring and Assessment, 50: 255-269.

Edwards, P., Kaewpaitoon, K., McCoy, E.W. and Chantachaeng, C. (1986). Pilot small-scale crop/livestock/fish integrated farm. AIT Research Report, 184, AIT, Bangkok, Thailand, 131 pp.

CHAPTER 13

IMPACT OF TEMPERATURE ON OUTBREAK OF EPIZOOTIC ULCERATIVE SYNDROME IN FISH

Tun-Tun Singh, Shambhu Prasad and N.K.Dubey

University Department of Zoology,
L.N.Mithila University, Darbhanga 846004
Bihar, India.

ABSTRACT

The present paper discuss about the influence of temperature on outbreak of epizootic ulcerative syndrome (EUS) in fish in association with seasonal variation. The study revealed that the gradual decline in temperature directly affect growth and survival of fish might be one of the important predisposing stress factor for out break of EUS in Fish.

INTRODUCTION

Aquaculture is the most promising and fast growing food fanning sector of India contributing 70% of total fish production and having an annual growth rate of 9 percent. Inadequate knowledge of fish health management among the fish farmers is one of the major constrants in popularizing large scale development of rural aquaculture entrepreneurship. Fish are always in equilibrium between potential disease organism and their environment. Change in equilibrium due to deterioration of water quality can lead to stress and consequently make fish prone to diseases (Boyd 1982 Boyd and Pillay - 1984). The outbreak of epizootic ulcerative syndrome in fish is one of the major threats and challenge to aquaculturists not only in India but throughout the world. The earlier work suggested heavy rainfall, decline in temperature and poor dissolved oxygen to be the important pre-disposing stressed factors for out break of epizootic ulcerative syndrome (Pathiratne, et.al. 1992, 2001. Lio-po-1999, Chinabutetal 1995).

MATERIALS & METHODS

The experiments were conducted during winter season (November 2002 to March 2003) to know the impact of temperature on outbreak of epizootic ulcerative syndrome (EUS). The experiments were conducted on a perennial pond having total water spread area of 5 has and average depth of 1.5 meter. The pond is situated near Baghmor of Darbhanga Town in Bihar state having 25.5' to 26.27' (North) latitude and 85.45'to 86.25' (East) longitude.

The species stocked in pond were Catla *(Catla Catla)*, Rohu *(Labeo, rohita)*, Mrigal *(Cirrhinus mrigala)* Grass carp *(Ctenophraryangodon idella)*, Common carp *(Cyprynus carpio)*, Silver carp *(Hypophthalmicthys molitrix)* and Calbasu *(Labeo calbasu)* having stocking density @ 10000 fingerlings per hectare . The fish were fed ricebran and mustared oil cake @ 2 to 5 percent of their body weight. The seed were stocked in the month of June- July 2002. The water temperature and dissolved oxygen value at the time of stocking were 28°C and 6.8 ppm respectively. The duration of experiment were fixed as fifteen week (i.e. 19 Nov. 2002 to 3 March 2003). The temperature was recorded every day in the morning and evening with the help of mercury thermometer and then average weekly temperature was calculated. The dissolved oxygen was also calculated every week throughout the investigation period by Winkler' method following standard method of APHA (1989). The record of the number of infected fish and mortality were maintained throughout the investigation period.

RESULTS & DISCUSSION

The average weekly fluctuation of temperature and dissolved oxygen (DO) along with average infected fish and average mortality recorded during 19 November 02 to 3 March 03 have been shown in table-13.1, and fig. 13.1 to 13.4. It was observed that DO decreases with the decline in temperature. However, the minimum temperature 9.4 and minimum DO 3.6 ppm were recorded in 9th week (i.e. 15 to 21 January 03). The result revealed positive correlation between temperature and DO.

The experiment was conducted in winter (November 2002 to March 2003) the water temperature declined significantly upto 15 January 2003 and afterward started rising gradually. It was observed that the number of infected fish and consequently the mortality increased with the gradual decrease of temperature and DO. When temperature declined from November 2002 the number of infected fish and mortality increased significantly and maximum no. of infected fish (170) and mortality(151) were recorded in the month of January. The average infection and the average mortality increased with the decease in temperature.

The Physico- chemical parameter like temperature and dissolved oxygen were receded through out the investigation period revealed that there is direct correlation between temperature and DO. During the investigation period the maximum temperature 22.9° C and DO 6.8 ppm recorded in 15th week (i.e. 26 February 03 to March 03 followed by temperature 21.45° C and DO 6.5 ppm in 1st week (19-25 November 02) and the minimum temperature 9.4° C and DO 3.6 ppm were found in 9th week (15-21 January 03) after that when temperature rise gradually the no. of infected fish and mortality decrease considerably. The result of present investigation revealed that the fluctuation of temperature and variation of DO might be one of the predisposing stress factor for outbreak of EUS. The similar observation were also reported earlier by Pathiratne *et al.*, (2001), Lio- Po G.D (1999) and Chinabut *et al.*, (1995) by monitoring the environmental influence on the out break of EUS in fresh water fish.

Thus the overall study and data might be suggestive of the fact that proper monitoring of Physico- chemical parameters of water followed their remedial measure is essential to prevent the out break of EUS in fish.

Table 13.1

Average Weekly Fluctuation of Temperature and Dissolved Oxygen in Association with average infected fish and average Mortality recorded during 19 Nov. 02 to 3 March 03.

No of week	Average Temperature (°c)	Dissolved Oxygen (in ppm)	Average Infected Fish (No)	Average Mortality (No)
1st(19-25Nov'02)	21.45	6.5	50	32
2nd (26 Nov-2 Dec '02)	20.8	6.5	58	36
3rd (3-9 Dec '02)	18.5	6.3	62	40
4th (10-16 Dec '02)	18.05	6.1	70	41
5th (17-23 Dec '02)	16.0	5.8	76	52
6th (24-31 Dec '02)	14.7	5.6	75	58
7th (1-7 Jan '03)	13.85	5.3	74	63
8th(8-14Jan'03)	9.8	4.1	150	122
9th (15-21 Jan '03)	9.4	3.6	170	151
10th(22-28 Jan'03)	11.85	4.2	140	125
11th(29 Jan-4 Feb'03)	13.45	4.5	150	84
12th(5-11- Feb'03)	16.75	5.3	170	63
13th(12-18 Feb'03)	18.00	5.8	170	44
14th(19-25 Feb '03)	20.5	6.1	160	33
15th(26 Feb-3 March' 03)	22.9	6.8	150	20

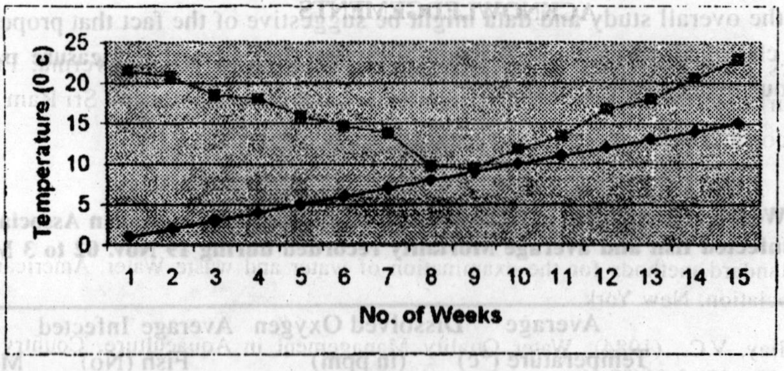

Fig. 13.1

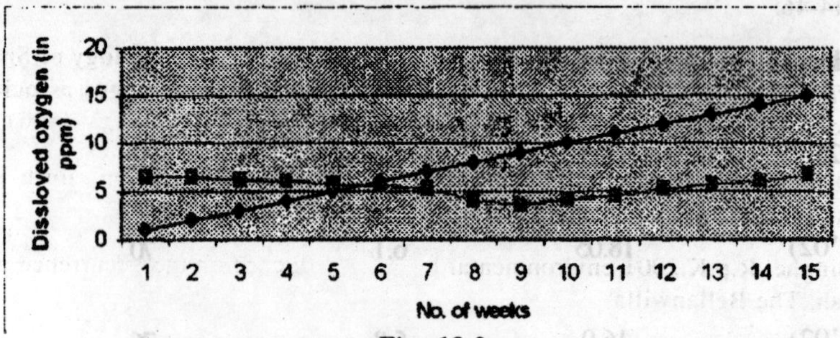

Fig. 13.2

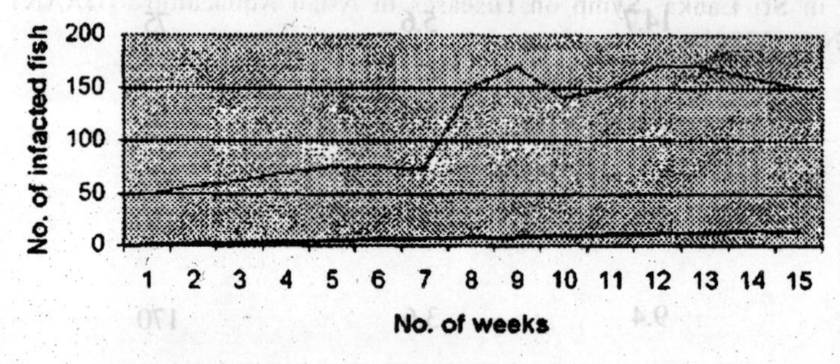

Fig. 13.3

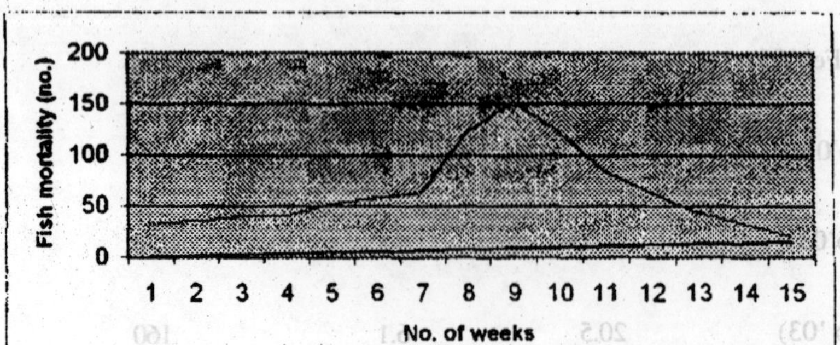

Fig. 13.4

ACKNOWLEDGEMENTS

The authors wish to express their gratitude to Dr. Shishir Kumar Verma, Principal Investigator, Department of Zoology, L.N.Mithila University, Darbhanga and Sri Ram Lakhan Sahani, for their co-operation and providing necessary facilities.

REFERENCES

APHA - (1985), Standard methods for the examination of water and waste Water, American Public Health Association, New York.

Boyd, CE, and Pillay, V.C., (1984), Water Quality Management in Aquaculture, Country Special Publication No. 22: 5-11/22.48.

Boyd, C.E. (1984) Water Quality Management for Pond Fish Culture, Elseviey Scientific Publishing Company P.44-46.

Chinabute, S., Roberts, R.J., Willoughby, U.R., Pearson, M.D., (1995). Histopathology of Snake Head *Channa straitus* (Bhoch), Experimentally infected with aphanomyces fungus associated with Epizootic Ulcerative Syndrome at different temperature, J. of fish diseases, V-18 (I) p. 91-97.

Lio-Po-G.D., (1999) Epizootic Alternative syndrome (EUS) recent development, forth symp. on disease in Asian Aquaculture (DAAN), Cebu City Phillipines; 22-26NOV.

Pathiratune, A., Jaysinghe. R.p.K.2001 environmental factures influence on the occurrence of EUS in freshwater fish. The Bellanwilla

Attidiya Wetland, Sri Lanka. J. of applied Ichthyology Vol. 17 (1) P. 30-34. Pathiratne A. Jaysinghe, R.P.K., and Epa, U.P.K. 1999 Environmental factors Associated with the occurance of EUS in freshwater fishes in Sri Lanka. Symp on Diseases in Asian Aquaculture (DAAN) Cevucity Phillipins 22-26 Nov. 1999.

CHAPTER 14

NEED FOR MULTIMEDIA STRATEGY FOR ... EXTENSION

ABSTRACT

The innovation in scientific technology ...

clientele. ...

all ...

transmitted through a variety ...

farmer frontier ...

process, the fishery extension personnel ...

The differential ...

multimedia for effective communication for ...

depicted that three media combination viz. ...

contributed the most in terms of overall effect ...

the technology of ...

West Bengal. ...

the increase in ...

inland fishery extension personnel to adopt ...

fish production in the country.

Key words: ...

INTRODUCTION

The ... inland fisheries ...

innovations in ... sector ...

NEED FOR MULTI-MEDIA STRATEGY FOR ACCELERATING INLAND FISH PRODUCTION

Utpal Bhaumik, T. Paria, S. K. Saha and A. Mitra

Central Inland Fisheries Research Institute, Barrackpore, Kolkala, 700120, West Bengal

ABSTRACT

The innovation of scientific technologies in inland fisheries sector not only has rendered support to increase fish production but also has imparted new dynamism to Indian fisheries and new hopes and confidence to million of fish farmers/fishermen. This has provided both a challenge and opportunity to the communication of information to the clientele. Experience in communication research showed that no single medium holds good in all situations. Research studies in communication indicated that the same message if transmitted through a variety of channels/media would leave and imprint on the memory of a farmer. In order to reach the fish farmers/fishermen at different stages of innovation decision process, the fishery extension personnel are required to use several media in combination. The differential sources credibility accorded by fish farmers demand the need to use multimedia for effective communication for technology. The study conducted at CIFRI depicted that three media combination *viz.* Group discussion + Slide show + Demonstration contributed the most in terms of overall effectiveness of communication while disseminating the technology of *Pen and Cage culture* of fish and prawn in the floodplain wetlands of West Bengal. Significant increase in the adoption of technologies has been observed with the increase in number of combination of media. Thus, concerted effort need be taken by the inland fishery extension personnel to adopt proper combination media for accelerating inland fish production in the country.

Key words : *Communication; Clientele: Message; Media; Dissemination*

INTRODUCTION

The breakthrough in inland fisheries production triggered off by the new scientific innovations in inland fisheries sector and their use on commercial scale particularly adoption of the high productive technologies and strategy has imparted a new dynamism to Indian fisheries and has given new hopes and confidence to millions fish farmers/fishermen. The Indian fisheries sector has heaved itself out of the ruts of traditional inland fish farming based on customs and traditions and is increasingly assimilating the modern and most scientific techniques for stepping up the productivity. This whole process of transformation of traditional inland fish farming into a modern Commercial enterprise has given strong support towards *Blue revolution.* All these have created unprecedented upsurge in the demand for appropriate, authentic and timely information on new technologies in inland fishery sector including effective use of fish health management, environmental impact assessment, fish conservation, inputs requirement and post-harvest technology and above all information on cost benefit ratio. This has provided both a challenge and an opportunity to the communication of information on inland fishery specially the research and extension works to the needs of rapidly changing and developing Indian fisheries and fish farming community. Experience in communication research showed that no single media holds good in all situations. The complexity of technology, various categories of farmers i.e., illiterate to highly educated and the variance in communication skills of extension personnel demand the need to have a combination of media for effective communication of technology. To find out the effectiveness of different combinations of media a study was conducted in 3 flood plain lakes of West Bengal during dissemination of pen and cage culture of fish and prawn.

MATERIAL AND METHODS

The investigation was carried out amongst the member fishermen of 3 Fishermen Cooperative Societies managing Saguna & Mathura and Amda beels of North 24 Parganas and Nadia districts of West Bengal respectively. Initially list of member fishermen of above 3 Fishermen Cooperative Societies were prepared and 25 members from each society were selected at random which constituted the sample of the study. Six different communication media viz. field trip, group discussion, demonstration, fish farmers' day, film sow, poster exhibition were utilized individually and in different combinations during dissemination of pen and cage culture operation of fish and prawn to assess their effectiveness. The effectiveness was judged on the basis of fishermen's score given to the extension techniques.

RESULT AND DISCUSSION

Communication researches on source credibility at CIFRI conclusively showed that different categories of fish farmers/fishermen attached differential credibility to various sources of information. Bhaumic and Saha (1998) reported that demonstration was ranked as Ist in sources credibility in rural areas whereas 3rd rank in suburban areas. Thus, the differential sources credibility accorded by fish farmers demand the need to use multimedia for effective communication of technology.

The modern inland fishery technologies with its complex nature demand a variety of media for effective dissemination of information to the fish farmers/fishermen. Bhaumik and Saha (1998) highlighted the need of using different communication media by the fishery extension personnel on the basis of the nature of technology.

Effectiveness of different combinations of media

The media commonly used for inland fisheries extension work are: Fieldtrip, group discussion, demonstration, films exhibition, radio, fish farmers' day, poster show etc. The relative effectiveness of each medium of communication has been experimentally found to differ.

It is clear from the Table 14.1 that Field trip was ranked as First in terms of overall effectiveness followed by Demonstration. On the other hand poster had lowest scores of in terms of awareness, knowledge and adoption. Chin-wen chen (1981) reported a relative influence of extension methods, personal contact (38.23%) was ranked as first in terms of number of farm families adopted the innovation. It was followed by indirect influence (25.11%),Group discussion (21.95%), Mass media (9.24%) and other influences (5.47%).

Table 14.1
Effectiveness of single medium

Rank in descending order of overall effectiveness	Media	Score
1	Field trip	7.32
2	Demonstration	6.67
3	Fish farmers' day	4.66
4	Group discussion	3.05
6	Film/show	2.82
7	Poster/exhibition	0.62

EFFECTIVENESS OF COMBINATION OF TWO MEDIA

Two media combinations were found to be statistically similar in effect. The combinations of poster + slide show and group discussion + slide show were ranked lowest in terms of overall effectiveness. Other media combinations are indicated in Table 14.2 in descending order of effectiveness. While studying the two media combinations. Singh and Singh (1976) reported that the combination of Field trip + Slide show emerged as the best in terms of overall effectiveness.

Table 14.2
Effectiveness of combinations of two media.

Rank in descending order of overall effectiveness	Media	Score
1	2	3
1.	Demonstration + Slide show	12.45
2	Demonstration + Group discussion	10.62
3.	Field trip + Group discussion	08.06
4.	Exhibition + Group discussion	07.18

1	2	3
5.	Radio + Demonstration	06.90
6.	Film + Group discussion	06.62
7.	Poster exhibition +Slide show	05.02
8.	Group discussion + Slide show	04.13

Mishra (1970) observed that the result and method demonstrations together have influenced 64 percent of farmers in changing their attitudes towards adoption of improved practices. Jalihal (1965) found that the effectiveness of the village meeting was increased due use of movie films. From the study mentioned above, it is clear that demonstration +Slide show or Demonstration+ Group discussion are the best two media combinations for motivate the fish farmers/fishermen to adopt the new technology.

EFFECTIVENESS OF COMBINATIONS OF THREE MEDIA

Among the three media combinations demonstration along with other two combinations contributed the most in terms of overall effectiveness as reported in Table 14.3. The combination of Group discussion + Slide show + demonstration, Flip chart + Group discussion + demonstration and Field trip + Group discussion + demonstration were found to be statistically similar to the top ranking combination to change the attitude of the fish farmer/ fishermen towards adoption of Pen and Cage culture technology in open water system. However, Mankar (1966) found that the most effective combination in brining about a change in knowledge, attitude and skill of farmers was personal contact method demonstration + leaflet. Rao (1958) observed that to popularise most improved agricultural practices the combination of Field trip +Movie + Result demonstration was the best.

Table 14.3 Effectiveness of combinations of three media

Rank in descending order of overall effectiveness	Media	Score
1	Group discussion + Slide show + demonstration	16.14
1	Poster exhibition + Group discussion + demonstration	15.62
2.	Field trip + Group discussion + demonstration	14.46
3.	Fish farmers' day + Group discussion + demonstration	12.76
4.	Field trip + Slide show + Group discussion	9.07
5.	Poster exhibition + Group discussion + Slide show	8.70
6.	Field trip + Group discussion + Slide show	7.48

CONCLUSION

No single communication medium is suitable under any situation of transfer of technology system. The proper combination of media, which increases the effectiveness of communication, is always warranted. In addition to this, the probability or changes of reaching message to the target: ownership of information by certain individuals, the complexity of technology, stage of the innovation decision process of an individual fish farmer/fishermen and differential credibility accorded by fish farmers to various sources of information, demand the need to have multi-media for effective transfer of technology. Significant increase in the adoption of inland fisheries technologies has been observed with the increase in number of combination media used. Thus, concerted effort need be taken by the inland fisheries extension personnel to adopt proper combination of media for accelerating inland fish production in the country.

ACKNOWLEDGEMENTS

The authors are grateful to the Director of CIFRI, for providing facilities for the research. Authors also thankful the Fishermen Cooperative Societies of North 24 Parganas and Nadia districts of West Bengal for rendering help for these study.

REFERENCES

Bhaumik, Utpal and Saha, S. K. 1998. Need for composite fish culture technology in West Bengal as perceived by the fish farmers. *Ed.* Thomas. P. C. *In* Current and Emerging trends in aquaculture. Daya Publshing House. New Delhi.

Chin-wen Chen, 1981. Effective extension approaches applied to Taiwan. Republic of Chaina. In extension education and rural development. Vol. 2. John Wiley & Sons Ltd.

Jalihal, K. A. 1965. The procedure followed in effecting a break through from demonstration to adoption stages in fertilizing in dry land Ragi in Mysore state. Alumni Association Bulletin. Agricultural College. Hebbal. Bangalore.

Mankar. S. S. 1966. Relative effectiveness of selected extension methods and their combinations in introducing chemical control of weeds. Unpublished Ph. D. Thesis.

University of Nagpur. Mishra. R. S. 1970. Quoted in agricultural communication at extension teaching methods including audio visuals aids. Sahay. B. N. Research in Extension Education of accelerating development process. Indian Agricultural Research Institute. New Delhi. 165- 183.

Rao, C. S. S. 1958. Evaluation of extension methods for successful introduction of improved farming practices. Ph. D. Thesis. Indian Agricultural Research Institute. New Dehli.

Singh. K. N. and Singh, S. N. 1976. Effective communication media for rural audiences (An experimental study). The Dharamsi Morarji Chemical corporation Ltd. 317/21. Dr. Dadabhai Naoroji Road, Bombay.

CHAPTER 15

Effect of Different Concentrations of Folidol on Glycogen in Liver of Snake Head Fish, *Channa Punctatus* (Bloch.)

Satyapal Sulodia and Surendra Singh

Department of Zoology, School of Life Sciences, Khandari Campus, Dr. B.R. Ambedkar University, Agra.

ABSTRACT

Pesticides are being used worldwide on large scale day by day. These are used in India 31930 tonnes during 1999 to 2001 according to Sunita Narian, 2001. These concentrations may sometimes reach to lethal levels causing fish mortality which has harmful effect on human being when consumed by them. The effect of Folidol at three different sub-lethal doses, 100 ppm, 200 ppm and 300 ppm was studied on total glycogen in full grown specimen of *Channa punctatus* (Bloch.).The glycogen contents was estimated. The total glycogen is significantly decreasing compared to control group. This change may be due to glycogen breakdown to glucose under the pesticide toxicity.

INTRODUCTION

Different kinds of pesticides are used to control pests. Pesticides have been used since about 1850, but before 1940s they were used in relatively small amounts and caused no harmful effects. Recently, a great deal of attention has been paid to evaluate the hazardous effect of organophosphorus compound on physiology of many non-target organisms, particularly fish. The pesticides pollute the aquatic ecosystem were they carried by rain water, from here they pass into the food chain, ultimately produced toxicity to fishes, birds, wild life and man. The present study evaluates the toxicity of different concentrations of a common pesticide, Folidol on glycogen in snake head fish *Channa punctatus* (Bloch.).

MATERIALS AND METHODS

Full grown specimen of *Channa punctatus* (Bloch.) also known as snake head fish, measuring 12-15 cm in length and 29-65 gm in weight, were selected for the experiments. They were acclimatized at the room temperature in glass aquaria (75 cm X 37.5 cm X 37.5 cm) containing 25 liters non-chlorinated tap water to assess the effect of Folidol the fish, *Channa punctatus* (Bloch.).

The LC_{50} value has been calculated by log dose/probit regression line method (Finney, 1971). The glycogen content was estimated by the method of Rex-Montgommery (1957).

OBSERVATIONS

The liver total glycogen in control group ranged from 0.024 to 0.032 with an average of 0.004 mg/dl. (Table 15.1)

Table 15.1

Effect of Folidol Toxicity On Liver Total Glycogen of *Channa punctatus* (Bloch.)

S.No	No.of observation	Exposure time in hours	Control		Experimental Set					
					100 ppm		200 ppm		300 ppm	
			Range	Mean±S.Em (mg/dl)	Range	Mean±S.Em (mg/gl)	Range	Mean±S.Em (mg/gl)	Range	Mean±S.Em (mg/gl)
1	3	24	.024-.032	.028±.002	.004-.036	.017±.0096*	.02-.028	.024±.002*	.032-.216	.025±.004*
2	3	48	.024-.032	.028±.002	.0016-.024	.016±.007*	.02-.016	$.016±.002^{\$}$	.002-.024	.0092±.007**
3	3	72	.024-.032	.028±.002	.004-.028	$.016±.006^{\#}$	.0016-.038	$.012±.0019^{\#}$	.0008-.012	$.0048±.0035^{\#}$
4	3	96	.024-.032	.028±.002	.0016-.032	$.012±.0019^{\#}$	.004-.012	$.008±.0023^{\#}$	.0004-.002	$.0017±.0007^{\#}$

ppm=parts per million
S.Em=Standard Error of Mean

* Non significant ($p>0.05$)
** Significant ($p<0.05$)
$^{\$}$Highly significant ($p<0.01$)
$^{\#}$Very highly significant ($p<0.0001$)

The liver total glycogen after intoxication of folidol at 100 ppm for 24 hours ranges from 0.004 to 0.036 with an average of 0.017 mg/dl, after 48 hours ranges from 0.0016 to 0.024 with an average of 0.016 mg/dl, after 72 hours ranged from 0.004 to 0.028 with an average of 0.016 mg/dl, while after at 96 hours ranges from 0.0016 to 0.033 with an average 0.012 mg/dl; at 200 ppm for 24 hours ranges from 0.02 to 0.028 with an average of 0.024 mg/dl, while after 48 hours ranges from 0.02 to 0.016 with an average of 0.016 mg/dl, while after 72 hours ranges from 0.0016 to 0.032 with an average of 0.012 mg/dl, while at 96 hours ranges from 0.004 to 0.012 with an average 0.008 mg/dl. At 300 ppm for 24 hours ranges from 0.032 to 0.216 with an average of 0.025 mg/dl, while 48 hours ranges from .0.002 to 0.024 with an average 0.0092 mg/dl, while 72 hours ranges from 0.0008 to 0.012 with an average 0.0049 mg/dl, while after 96 hours ranges 0.0004 to 0.002 with an average 0.0017 mg/dl.

The decrease in liver total glycogen of the exposure time of folidol, compared to control is non significant (p>0.05) at the 24, 48 and 72 hours but very highly significant (p0.00l) at 96 hours at the 100 ppm of sub-lethal concentration, while 200 ppm of sub-lethal concentration non significant (p>0.05) at 24 hours and highly significant at 48 hours and very highly significant (p<0.00l) at 72 and 96 hours, while at 300 ppm of sub-lethal concentration at 24 hours non significant (p>0.05) at 48 hours significant (p<0.05), 72 and 96 hours is very highly significant (p<0.001) was observed.

RESULTS AND DISCUSSION

The liver glycogen shows decreasing trend on exposure to Folidol at different tome intervals (24, 48, 72 and 96 hrs) at 100 ppm sub-lethal concentration. The similar trend was observed at 200 ppm and 300 ppm sub-lethal concentration. The fall in glycogen content in the tissue indicates its rapid utilization by the respective tissue as a consequence of pesticide toxic stress due to glycogen break down to glucose.

Present findings gain support with the findings of Cope *et al.* (1970), similar decrease in tissue glycogen content of liver was reported in *Tilapia mossambica* treated with malathion by Kabeer Ahmad *et al.* (1980), since glycogen depletion is more prevented under hypoxia conditions, Dezwaan and Zandee (1972) and Devi (1982).

REFERENCES

Cope, O.B. *et al.* 1970. Some chronic effects of 2,4-D on blue gills, *Laponis macrochurus,* Trans. Am. Fish. Soc., 99:1.

Devi, A.P. *et al.* 1982. The effect of endosulfan and its isomer on tissue protein, glycogen and lipids in the fish *Channa punctatus.* Pestic. Biochem. Physiol., 17(3):282-296.

Dezwaan, A. and Zandee, D.I. 1972. The utilization of glycogen and accumulation of some intermediate during anaerobiosis in *Mytius edulis* L. Comp. Biochem. Physiol. B 43, 47-54.

Kabeer, A.S. and K.V. Ramana Rao. 1980. Toxicity of malathion to the fresh water fish, *Tilapia mossambica.* Bull. Environ. Contain. Toxicol., 24:870-874.

Montgommery, R. 1957. The determination of glycogen. Arch. Bioch. Biophys., 67 : 378-386.

Narian, S.2001 director center for science and environment.

CHAPTER 16

EFFECT OF SIZE AND DENSITY ON THE FEEDING ENERGETICS OF THREE FRESHWATER FISHES

A. Jeyaseeli* and M. Narayanan

Aquatic Biodiversity Centre, St. Xavier's College (Autonomous), Palayamkotti - 627 002, India

**Department of Zoology, Jayaraj Annapackiam College for Women, Periyakulam - 625 601, India*

ABSTRACT

In the present study Cr, Ar, Mr and Pr for the different size gorups and density series were observed in *H. fossilis, M. keletius* and *C. carpio* respetively. In all the tested fishes size and stocking density significantly ($P<0.05$) affected the feeding energetics parameters and these parameters were inversely related with the size groups. The oberved absorption efficiency of the experimental fishes varied between 88.00 to 94.90%.

INTRODUCTION

Food is one of the important factors significantly affecting the fish growth (Farooq — Anwar and Jafri, 1995 and Firdaus & Jafri, 1996). When fishes were allowed to feed on amimal food it resulted in a better conversion of food into body substance (Narayanana, 1980 and Arunachalam *et al.*, 1985). The factors that determining the economic feasibility fo cultivable fish is their capacity to produce more, even when they are stocked at a high density. High density culture might affect fish production through the reduction in the availability fo space and food. Oxygen depletion and accumulation of various metabolities are the major problems associated with high density culture. Studies on certain species of fish have revealed the negative effects of stocking density on growth rate under intensive rearing (Holm *et. al.*, 1990 and Sakthivel and Sampath, 1992); a few have reported the absence of negative effect (Watanabe *et al.*, 1990 and Martines - Tapia and Fernandez — Pato, 1990).

Stocking density is an important parameter which directly affects the growth of fish and hence its production in fish population (Backiel & Lecren, 1978). The use of high stocking density as a technique to maximize water usage and thus increase fish production has been shown to exert severe adverse effects on growth and survival rates (Jhingran, 1991 and Jena *et. al.*, 1998). Therefore, it is necessary to predetermine and standardise the optimum stocking density for each species in order to obtain the best possible output. Though high stocking density is considered a stressor by many researchers (Ainsworth *et al.*, 1985; and *Gatlin et al.*, 1986) little progress is made to identify how high stocking density influences the growth rate of fish. Therefore, the present investigation was carried out to compare the influence of density and size on the food utilization in three freshwater fishes namely *Heteropuenstes fossilis, Mystus keletius* and *Cyprinus carpio.*

MATERIALS AND METHODS

The experimental animals *H. fossilis, M. keletius* and *C. carpio* were collected from ponds in and around Palayankottai. They were brought to the laboratory and stocked in large rectangular cement tanks (500 1 capacity) for a period of 10 to 15 days in well aerated water. During the period of acclimation they were fed daily with chopped beef liver (experimental food). These fishes were then sorted into three size groups namely A, B and C weighing 4.46 ±0.148, 7.79 ±0.290 and 12.11 ± 0.116 g of *H. fossilis,* K, L and M weighing 4.23 ± 0.125, 7.77 ± 0.306 and 12.04 ± 0.126 g of *M. keletius* and X, Y and Z weighing 4.36 ± 0.412, 7.79 ± 0.386 and 12.05 ± 0.115g of *C.carpio* respectively. The water was changed once daily during the period of acclimation in order to keep the dissolved oxygen content above 50% saturation level.

Fish of the respective size groups *of H. fossilis, M. keletius* and *C. carpio* were grouped into different densities in logarithmic order *viz.*, 1, 2, 4, 8 and 16 fish / aquarium (151 capacity). Each size group thus contained five densityseries. However, in larger size groups (B&C, L&M and Y&Z), the 16 - fish series was not experimented with, since the capacity of the experimental troughs was found to be inadequate. Experimental animals were fed *ad libitum* with beef liver and the experiments lasted for 21 days. All the experiments were conducted at room temperature (29 ± 1°C) under normal photoperiod. The 'Sacrifice method' of Maynard and Loosli (1962) was followed in the present investigation to estimate fish growth. The results were statistically analysed following Zar (1984).

The scheme of energy balance followed in the present study is the IPB formula of Petrusewicz and Macfadyen (1970) which is usually represented as,

$$C=P+R+F+U$$

where C is the energy consumed, P is the growth (conversion), R is the energy lost as heat due to metabolism and F is the faeces.

$$\text{Consumption rate (Cr) KJ/g/day} = \frac{\text{Energy consumed (KJ)}}{\text{Initial wet weight x 21 days}}$$

$$\text{Absorption rate (Ar) KJ/g/day} = \frac{\text{Energy absorbed (KJ)}}{\text{Initial wet weight x 21 days}}$$

$$\text{Conversion rate (Pr) = KJ/g/day} = \frac{\text{Energy convered (KJ)}}{\text{Initial wet weight x 21 days}}$$

$$\text{Metabolic rate (Mr) = KJ/g/day} = \frac{\text{Energy metabolized (KJ)}}{\text{Initial wet weight x 21 days}}$$

RESULTS AND DISCUSSIONS

Effect of Size

In all the size groups tested, the smaller size group showed the maximum feeding rate and thereafter the rate decreased as the size increased in all the three tested species of fish (Tables 16.1, 16.2 and 16.3). Higher absorption and metabolic rates were observed in smaller size groups over that of larger size groups. This difference in the smaller size groups might be due to a higher energy demand including swimming activity (Warren and Davies, 1967). In the present work the mean absorption efficiency of various size groups of *H. fossilis, M. keletius* and *C. carpio* were 90.25, 92.84 and 94.29% respectively. Vivekanandan (1977) obtained the absorption efficiency for different weight classes of *Ophiocephalus striatus* averaged to 96.10%. Similarly Menzel (1960), Pandian (1967) and Gerking (1971) also reported that absorption efficiency hardly vary with weight classes. Hence it may concluded that the body weight did not appreciably alter absorption efficiency.

Table 16.1 : Effect of Density on the Feeding Energetics Parameters in Different Size Groups of *H. fossilis* Fed with Beef Liver for a Period of 21 days. Each Value Represents Mean ±SD of Three Replications

Size Group	Density	Initial wet wt. (g / animal)	Rate of energy consumption (KJ/g/day)	Rate of energy Absorption (KJ/g/day)	Rate of energy Metabolished (KJ/g/day)	Rate of energy Conversion (KJ/g/day)	Absorption efficiency (%)	Gross Conversion Efficiency K_1 (%)	Net Conversion efficiency K_2 (%)
	1	4.40±0.200	0.21±0.013^{a}	0.19±0.012^{a}	0.17±0.008^{ab}	0.02±0.003^{a}	91.16±0.490	12.13±1.086^{a}	13.29±1.140^{a}
	2	4 40±0 050	0 25±0.006^{b}	0.23±0.011^{b}	0.19±0016^{b}	0.04±0.006^{b}	90.90±0.430	15.87±3.070^{ab}	17.46±3.460^{ab}
A	4	4.56±0.104	0.26±0.007^{b}	0.24±0.008^{bc}	0.20±0.016^{b}	0.04±0.008^{b}	90.12±0.500	17.23±3.030^{ab}	19.10±3.300^{ab}
	8	4.55±0.020	0.29±0.007^{c}	0.25±0.008^{c}	0.21±0.004^{b}	0.05±0.004^{c}	89.49±0.970	18.83±1.350^{b}	21.05±1.730^{b}
	16	4.51±0.030	0.24±0.015^{a}	0.19±0.010^{a}	0.17±0.014^{a}	0.02±0.004^{a}	90.36±0.305	12.41±2.560^{a}	13.73±2.880^{a}
	1	8.20±0.100	0.16±0.004^{a}	0.14±0.003^{a}	0.12±0.007^{a}	0.02±0.004^{a}	88.50±0.780	14.56±2.800^{a}	16.45±3.220^{a}
B	2	8.33±0.104	0.19±0.005^{b}	0.17±0.005^{a}	0.14±0.001^{b}	0.03±0.003^{b}	89.20±0.500	17.57±1.340^{a}	19.67±1.530^{a}
	4	8.33±0.160	0.23±0.004^{c}	0.21±0.003^{b}	0.17±0.002^{c}	0.04±0.004^{c}	90.70±0.110	18.07±1.570^{a}	19.91±1.760^{a}
	8	8.32±0.170	0.22±0.013^{d}	.0.20±0.012^{b}	0.17±0.013^{c}	0.03±0.006^{b}	91.28±0.180	16.34±2.680^{a}	17.89±2.900^{a}
	1	12.13±0.152	0.13±0.007^{a}	0.12±0.006^{a}	0.11±0.002^{ab}	0.01±0.004^{a}	91.28±0.180	13.75±2.300^{a}	17.89±2.900^{a}
C	2	12.15±0.050	0.17±0.008^{b}	0.15±0.008^{ab}	0.13±0.011^{b}	0.02±0.003^{ab}	89.30±1.900	13.59±2.900^{a}	15.02±3.280^{a}
	4	12.03±0.076	0.23±0.009^{c}	0.21±0.008^{b}	0.18±0.118^{C}	0.03±0.004^{b}	92.30±0.500	15.27±2.400^{a}	16.51±2.660^{a}
	8	12.07±0.167	0.14±0.010^{a}	0.13±0.010^{a}	0.11±0.010^{a}	0.02±0.004^{a}	88.73±0.900	15.62±2.780^{a}	17.60±3.200^{a}

Figures in the same column having the same superscript are not significantly diferent ($P>0.05$)

Table 16.2

Effect of Density on the Feeding Energetics Parameters in Different Size Groups of *M. Keletius* Fed with Beef Liver for a Period of 21 days. Each value Reprsents Mean ± SD of three replications.

Size Group	Density	Initial wet wt. (g / animal)	Rate of energy consumption (KJ/g/day)	Rate of energy Absorption (KJ/g/day)	Rate of energy Metabolished (KJ/g/day)	Rate of energy Conversion (KJ/g/day)	Absorption efficiency (%)	Gross Conversion Efficiency K_1 (%)	Net Conversion efficiency K_2 (%)
	1	4.16±0.152	0.19±0.024^{a}	0.17±0.015^{a}	0.15±0.013^{a}	0.02±0.003^{ab}	90.30±4.200	12.60±0.348^{ab}	13.97±0.510^{ab}
	2	4.35±0.100	0.22±0.026^{ab}	0.21±0.025^{b}	0.17±0.023^{ab}	0.03±0.003^{b}	93.09±0.500	14.80±0.249^{b}	15.89±0.353^{b}
K	4	4.39±0.036	0.25±0.018^{b}	0.23±0.013^{b}	0.19±0.006^{b}	0.01±0.006^{b}	1.80±1.600	15.49±1.330^{b}	16.87±1.720^{b}
	8	4.32±0.075	0.27±0.020^{b}	0.25±0.013^{b}	0.20±0.010^{b}	0.05±0.006^{b}	90.70±0.870	18.52±1.680^{b}	20.41±1.980^{b}
	16	4.38±0.071	0.17±0.022^{a}	0.16±0.020^{a}	0.15±0.032^{c}	0.01±0.008^{a}	93.04±0.120	6.28±3.630^{c}	6.75±3.920^{c}
	1	8.10±0.200	0.16±0.030^{a}	0.14±0.028^{ab}	0.12±0.023^{a}	0.02±0.005^{a}	90.60±0.520	11.41±0.960^{a}	12.87±0.900^{a}
	2	8.10±0.150	0.19±0.008^{ab}	0.17±0.008^{ab}	0.15±0.006^{a}	0.02±0.002^{a}	91.30±0.230.	12.003±0.870^{a}	13.13±0.930^{a}
L	4	8.17±0.025	0.21±0.010^{b}	0.19±0.010^{a}	0.16±0.006^{a}	0.03±0.004^{b}	91.08±0.180	15.42±1.400^{b}	16.91±1.510^{b}
	8	8.19±0.077	0.15±0.022^{a}	0.13±0.025^{b}	0.12±0.017^{a}	0.01±0.004^{a}	90.43±0.760	11.83±1.360^{a}	13.07±1.410^{a}
	1	11.90±0.100	0.13±0.009^{ac}	0.12±0.001^{a}	0.11±0.006^{a}	0.01±0.003^{a}	90.91±0.190	11.61±1.690 ^{a}	12.63±1.700^{a}
M	2	12.00±0.080	0.16±0.018^{a}	0.14±0.010^{a}	0.13±0.012^{a}	0.01±0.004^{a}	90.23±0.350	10.51±1.830^{a}	11.63±2010^{a}
	4	12.05±0.025	0.21±0.016^{b}	0.18±0.012^{b}	0.16±0.012^{a}	0.02±0.004^{b}	91.20±0.500	12.74±1.050^{a}	13.96±1.080^{a}
	8	12.02±0.019	0.13±0.010^{c}	0.12±0.010^{a}	0.11±0.006^{a}	0.01±0.002^{a}	91.33±0.650	10.65±2.020^{a}	11.66±2.300^{a}

Figures in the same column having the same superscript are not significantly different ($P>0.05$)

Table 16.3

Effect of Density on the Feeding Energetics Parameters in Different Size Groups of *C. carpio* Fed with Beef Liver for a Period of 21 Days. Each Value Represents Mean ± SD of three Replications.

Size Group	Density	Initial wet wt. (g / animal)	Rate of energy consumption (KJ/g/day)	Rate of energy Absorption (KJ/g/day)	Rate of energy Metabolished (KJ/g/day)	Rate of energy Conversion (KJ/g/day)	Absorption efficiency (%)	Gross Conversion Efficiency K_1 (%)	Net Conversion efficiency K_2 (%)
	1	4.20±0.200	0.98±0.126^{a}	0.93±0.120^{a}	0.73±0.109^{a}	0.19±0.119^{a}	94.90±0.260	20.21±1.469^{a}	21.30±1.563^{a}
	2	4.23±0.104	0.82±0.071^{ab}	0.78±0.071^{ab}	0.60±0.072^{a}	0.17±0.005^{ab}	94.60±0.360	21.30±2.229^{a}	22.51±2.440^{a}
X	4	4.44±0.052	0.81±0.050^{ab}	0.77±0.046^{ab}	0.60±0.043^{a}	0.16±0.008^{bc}	94.60±0.152	20.51±1.190^{a}	21.65±1.260^{a}
	8	4.37±0.019	0.78±0.059^{ab}	0.74±0.056^{ab}	0.58±0.052^{a}	0.15±0.011^{bc}	94.00±0.500	19.73±0.007^{a}	20.84±0.108^{a}
	16	4.28±0.027	0.70±0.065^{b}	0.66±0.062^{b}	0.52±0.071^{a}	0.14±0.015^{c}	93.80±0.230	20.60±3.398^{a}	21.95±3.670^{a}
	1	8.10±0.200	0.58±0.030^{a}	0.55±0.036^{a}	0.43±0.034^{a}	0.13±0.012^{a}	94.80±0.300	23.93±3.608^{a}	25.23±3.870^{a}
Y	2	8.08±0.125	0.55±0.029^{a}	0.53±0.020^{a}	0.42±0.033^{a}	0.12±0.009^{a}	94.90±0.150	22.55±2.920^{a}	23.10±1.940^{a}
	4	8.14±0.052	0.52±0.052^{a}	0.49±0.051^{a}	0.37±0.040^{a}	0.12±0.012^{a}	94.70±0.305	23.21±1.270^{3}	24.69±1.440^{a}
	8	8.11±0.101	0.32±0.020^{b}	0.30±0.022^{b}	0.21±0.043^{b}	0.07±0.011^{b}	93.50±0.600	22.84±3.140^{a}	24.22±3.480^{a}
	1	11.96±0.050	0.51±0.038^{a}	0.48±0.036^{a}	0.38±0.023^{a}	0.10±0.007^{a}	94.90±0.100	21.09±1.490^{2}	22.13±1.740^{a}
Z	2	12.15±0.180	0.47±0.050^{a}	0.44±0.048^{a}	0.34±0.055^{ab}	0.10±0.008^{a}	94.50±0.208	22.00±4.010^{a}	23.26±4.300^{a}
	4	12.01±0.038	0.42±0.041^{ab}	0.40±0.038^{a}	0.30±0.032^{ab}	0.09±0.009^{a}	94.80±0.305	21.80±4.140^{a}	22.97±4.356^{a}
	8	12.01±0.116	0.30±0.027^{b}	0.28±0.029^{b}	0.22±0.026^{b}	0.05±0.016^{b}	91.80±2.830	18.14±7.080^{a}	20.01±7.490^{a}

Figures in the same column having the same superscript are not significantly different ($P>0.05$)

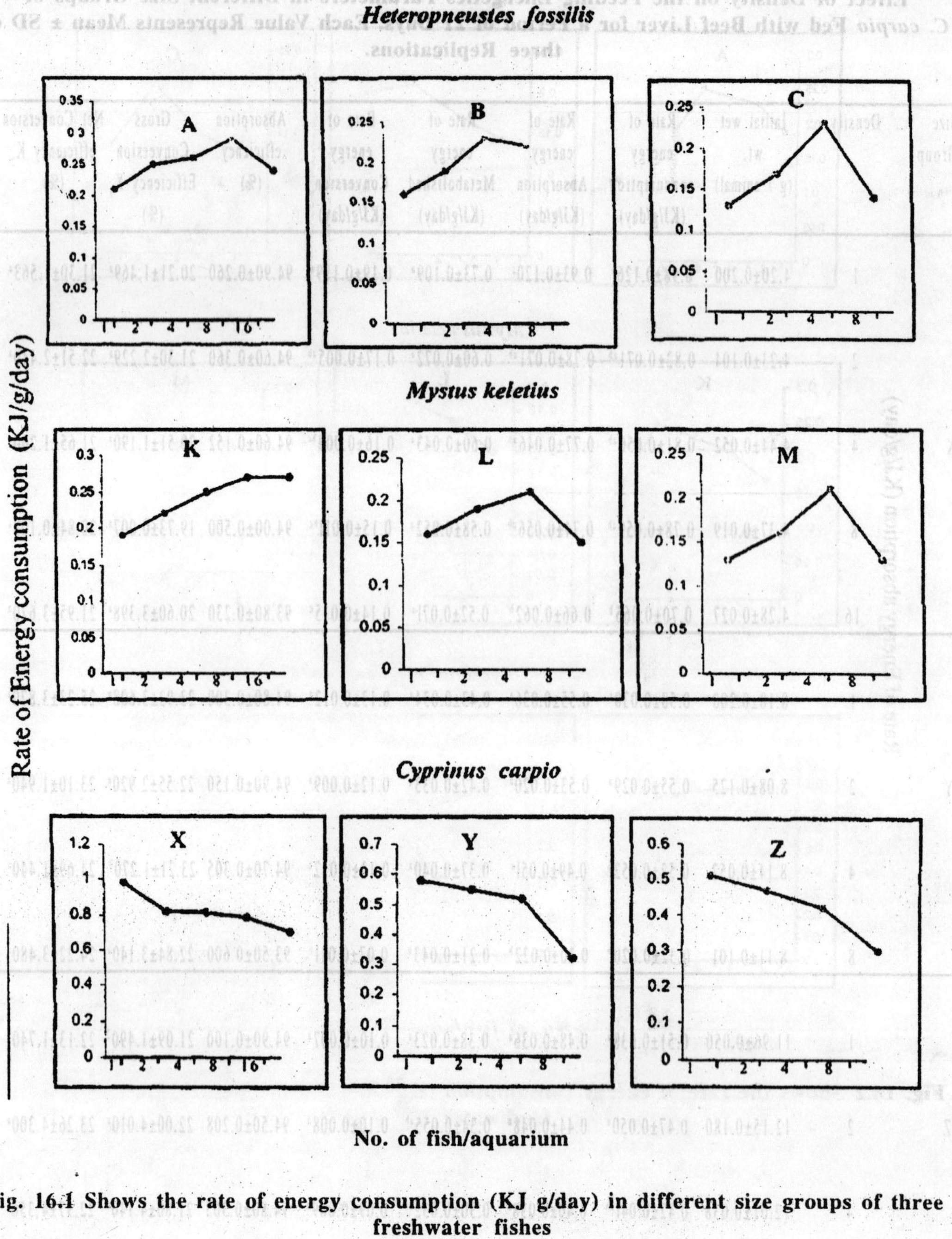

Fig. 16.1 Shows the rate of energy consumption (KJ g/day) in different size groups of three freshwater fishes

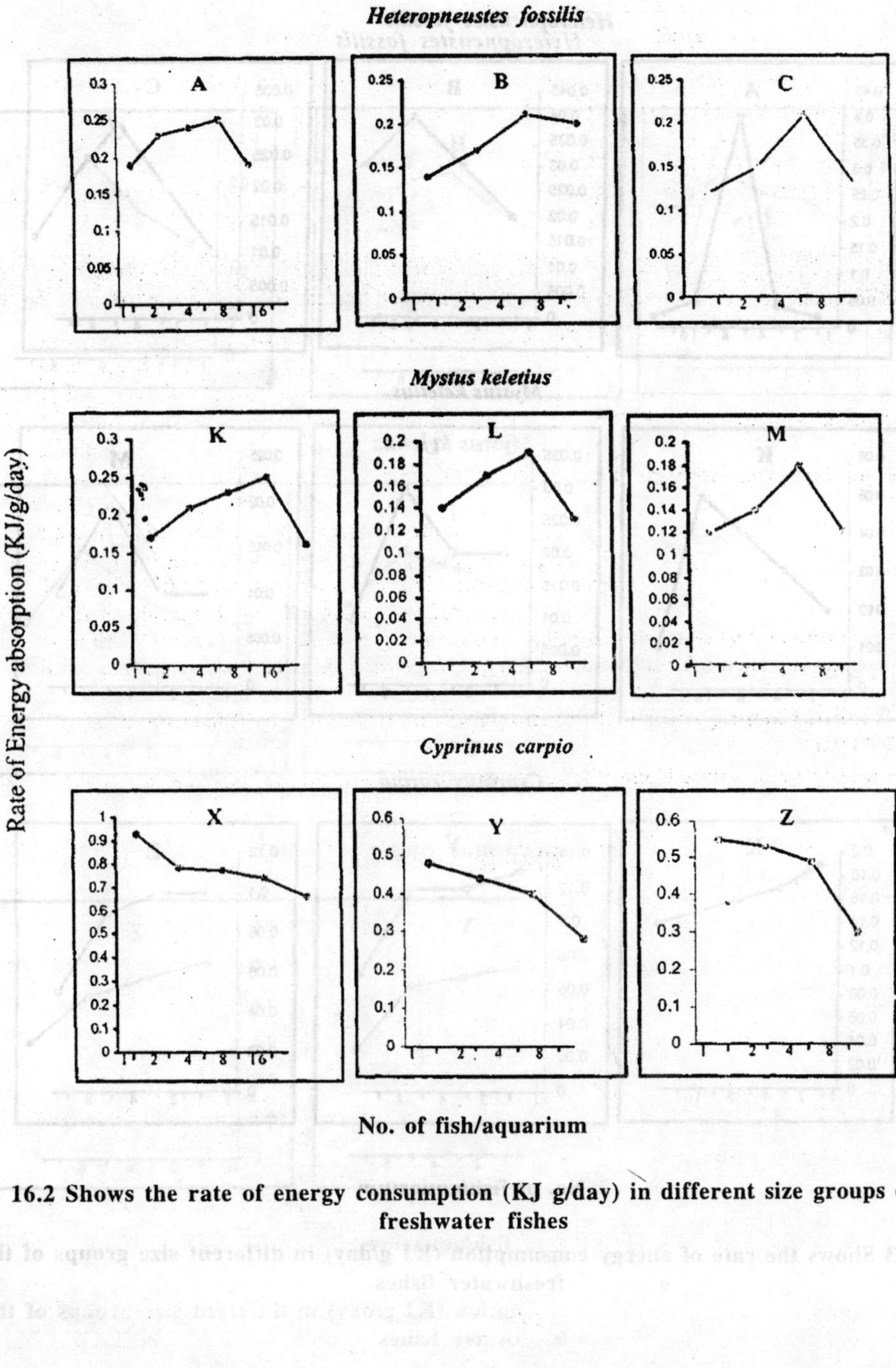

Fig. 16.2 Shows the rate of energy consumption (KJ g/day) in different size groups of three freshwater fishes

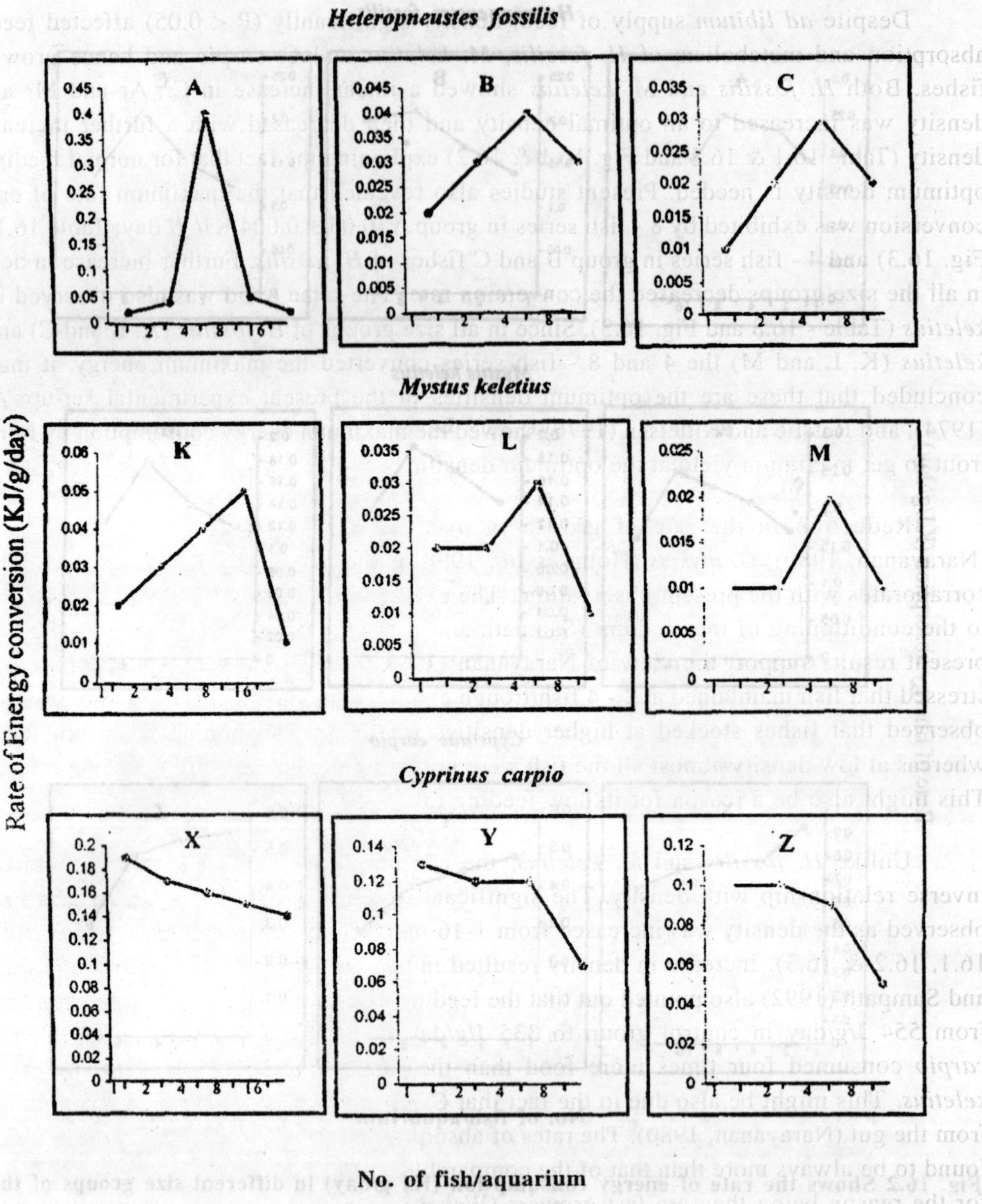

Fig. 16.3 Shows the rate of energy consumption (KJ g/day) in different size groups of three freshwater fishes

Effect of Density

Despite *ad libitum* supply of feed, density significantly ($P < 0.05$) affected feeding, absorption and metabolism of *H. fossilis, M. keletius* and *C. carpio* and hence growth of fishes. Both *H. fossilis* and M. *keletius* showed a linear increase in Cr, Ar and Mr as the density was increased to an optimal density and then decreased with a further increase in density (Table 16.1 & 16.2 and Fig. 16.1 & 16.2) explaining the fact that for normal feeding an optimum density is needed. Present studies also revealed that the maximum rate of energy conversion was exhibited by 8 - fish series in group A (0.05 ± 0.004 KJ/ g/ day; Table 16.1 and Fig. 16.3) and 4 - fish series in group B and C fishes of *H. fossilis.* Further increase in density in all the size groups decreased the conversion rate. The same trend was also observed in *M. keletius* (Table - 16.3 and Fig. 16.3). Since in all size groups of *H. fossilis* (A, B and C) and *M. keletius* (K, L and M) the 4 and 8 - fish series converted the maximum energy, it may be concluded that these are the optimum densities in the present experimental set-up. Allen (1974), and Refstie and Kittelsen (1976) showed the maximum energy consumption by rainbow trout to get maximum yield at the optimum densities.

Reduction in the rate of growth in over *crowded conditions* of *S. mossambicus* (Narayanan, 1980), *O. mykiss* (Holm *et. al.,* 1990) and *C.gariepinus* (Siddiqui *et. al.*, 1993) corraborates with the present observation.. The existence of the optimum density may be due to the conditioning of the medium (Sampath and Pandian, 1980 and Narayanan, 1980). The present results support the view of Narayanan (1980) who worked on *S. mossambicus* also stressed that fish maintained at 2 - 4 fish/trough exhibited maximum energy intake. It was also observed that fishes stocked at higher densities were found swimming when not feeding, whereas at low density almost all the fish were normally idle on the bottom when not feedmg. This might also be a reason for its low feeding rate at higher densities.

Unlike *H. fossilis* and *M. keletius,* the consumption rate of C. *carpio* exhibited an inverse relationship with density. The significant ($p<0.05$) reduction in Cr, Ar and Pr was observed as the density was increased from 1-16 individuals per trough (Table 16.3 and Fig. 16.1, 16.2 & 16.3). Increase in density resulted in decreased energy consumption. Sakthivel and Sampath (1992) also pointed out that the feeding rate of *C. carpio* decreased significantly from 554 J/g/day in control group to 335 J/g/day in the highest density group. *Cyprinus carpio* consumed four times more food than the same size groups *of H. fossilis* and *M. keletius.* This might be also due to the fact that *C. carpio* took less time to evacuate the food from the gut (Narayanan, 1980). The rates of absorption and metabolism of *C. carpio* was also found to be always more than that of the comparable size groups *of H. fossilis* and *M. keletius* for the reason being they are fast growers (Wahab *et al.,* 2002) and obtain naturally bigger sizes than the other two species. This negative correlation between growth rate and density was also observed by Teng and Chua (1978), Narayanan (1980) and Sakthivel and Sampath (1992). This might be due to the fact that growth depends on the food intake which inturn dependent on the population density. In the present study the reduced food conversion rate at high densities might be due to increased oxygen consumption and behavioural events (Li and Brocksen, 1977; Refstie, 1977 and Jobling, 1985).

The observed gross and net conversion efficiencies also agreed with the reports of the previous workers (Narayanan, 1980 and Arunachalam *et al.,* 1985). Invariably in the different densities of all the size groups *of H. fossilis, M. keletius, and C. carpio* the absorption efficiency varied between 88.00 to 94.90% (Tables 16.1, 16.2 and 16.3). These values coincided with the reports of Arunachalam *et al.*, (1985), Marian *et. al.*, (1982) and Sakthivel and Sampath (1989). Sampath and Pandian (1980) observed an increase in absorption efficiency from 90 to 96 % when the density was increased from 1-16 individuals /aquarium. However, such compensation by increased absorption efficiency for reduced food intake was not observed in these experimental fishes. Present analysis made us to conclude that the Cr, Ar and Pr of the three species were influenced by both size and density and aquaculturists should be careful about overcrowding, which is detrimental to production.

ACKNOWLEDGEMENTS

We wish to thank Prof. Thomas Punithan, H.O.D. and Dr. M.A. Haniffa, Director (CARE) for providing facilities. One of the authors (A.J.) wants to thank Rev. Sr. Alphonsal, JA College, Periakulam for her encouragement. This paper forms a part of the Ph.D thesis submitted by A.J. to M.S. University.

REFERENCES

Ainsworth, A.J., Bowser, P. R. and Beleau, M. H., (1985). Serum cortisol levels in channel catfish from production ponds. Prog. Fish - cult., 47 : 176 - 181.

Allen, K. O., (1974). Effect of stocking density and water exchange rate on growth and survival of channel catfish, *Ictalurus punctatus* (Refinesque), in circular tanks. Aquaculture, 4 : 29 - 39.

Arunachalam, S., V. Kalyansundaram and Palanichamy, S., (1985). Effects of food quality on food intake, growth and conversion efficiency in the air - breathing catfish, *Heteropneustes fossil is* (Bloch). Compatative Physiol. and Ecol., 10 (3) : 101 - 104.

Backiel, T. and Lecren, E. D., (1978). Some density relations on fish population parameters. In : S. D. Gerking (Editor), Ecology of fresh water fish production. Blackwell Scientific Publications, Oxford, 279-302.

Farooq - Anwar, M. and Jafri, A. K., (1995). Effect of varying dietary lipid levels on growth, feed conversion, nutrient retention and carcass composition of fingerling catfish, *Heteropneustes fossilis.* Asian Fish Sci., 8(1) : 55-62.

Firdaus, S and Jafri, A. K., (1996). Growth and haematological response of young catfish, *Heteropneustes fossilis* (Bloch) to dietary phosphorus supplementation. J. Aquacult. Trop., 11(4) : 263-269.

Gatlin, D. M., Poe, W. E., Wilson, R. P. Ainswoth, A. J. and Bowser, P. R., (1986). Effects of stocking density and vitamin C status on vitamin E - adequate and Vitamin E - deficient of fingerling channel catfish. Aquaculture, 56 : 187- 195.

Gerking, S.D., (1971). Influence of rate of feeding and body weight on protein metabolism of blue gill sunfish. Physiol. Zool., 28 : 9 - 19.

Holm, J. C., Refstie, T. and B, S., (1990). The effect of fish density and feeding regimes on individual growth rate and mortality in rainbow trout *(Oncorhynchus mykiss).* Aquaculture. 89 : 225 - 232.

Jena, J. K., Aravindakshan, P. K. and Singh, W. J., (1998). Nursery rearing of Indian major carp fry under different stocking densities. Indian J. Fish., 45 (2) : 163 - 168.

Jhingran, V. G., (1991). Fish and Fisheries of India (3rd edn.), Hindusthan Pub. Co., New Delhi, 727 .

Jobling, M., (1985). Physiological and social constraints on growth of fish with special reference to Artie charr, *Salvelinus alpinus* L. Aquaculture, 44 : 83 - 90.

Li, H. W and Brocksen, R. W., (1977). Approaches to the analysis of energetic costs of intraspecific competition for space by rainbow trout *(Salmo gairdnei)*. J. Fish. Biol., 11 : 329 - 341.

Marian, M. P., Ponnaiah, A. G. Pitchairaj, R. and Narayanan M., (1982). Effect of feeding frequency on surfacing activity and growth in the air - breathing fish, *Heteropneustes fossilis*. Aquaculture, 26: 237-244.

Martinez - Tapia C. and Fernandez - Pato, C. A., (1991). Influence of stocking density on turbot *(Scophthalmus maximus* L.) growth. ICES, C. M. F : 20.

Maynard, A.L. and Loosli, K.J., (1962). Animal Nutrition. 5th ed., McGraw-Hill, New York.

Menzel, D. W., (1960). Utilization of food by a Bermuda reef fish, *Epinephelus guttatus*. J. Con. Int. Explor. Men, 25 : 216-222.

Narayanan, M., (1980). Study of food utilization by multi - species laboratory population of fish. Ph. D. Thesis, Madurai Kamaraj University, India.

Pandian, T. J., (1967). Food intake, digestion, absorption and conversion in the fish, *Ophiocephalus striatus*. Helgol. Wiss. Meeresunters, 15: 637-647.

Petrusewicz, K. and Macfadyen, A., (1970). Productivity of terrestrial animals. IBP Hand Book, No. 13. Black Well Scientific Publications, Oxford.

Refstie, T., and Kittlesen, A., (1976). Effects of density on growth and survival of artificially reared Atlantic salmon. Aquaculture, 8:319- 326.

Refstie, T., (1977). Effect of density on growth and survival of rainbow trout. Aquaculture, 11:329-334.

Sakthivel, M. and Sampath, K., (1989). Effects of textile dye stuff effluent on food utilization of *Cyprinus carpio* (Linn.). Indian. J. Expt. Biol., 27 : 1032 - 1034.

Sakthivel, M. and Sampath, K., (1992). Effects of stocking density on growth in *Cyprinus carpio* (Linnaeus). Third Asian Fisheries Forum, Singapore, 149.

Sampath, K. and Pandian, T. J., (1980). Effects of density on food utilization and surfacing behaviour in the obligatory air-breathing fish, *Channa striatus*. Hydrobiologia, 68: 113-118.

Siddiqui, A. Q., A. R. Alnajada and Alhinty, H. M., (1993). Effects of density on growth, survival and yield of African catfish, *Clarias gariepinus* (Burchell) in Saudi Arabia. J. Aqua. Trop., 8:213-219.

Teng, S. K. and Chua, T. E., (1978). Effect of stocking density on the growth of estuary grouper, *Epinephelus salmoides* (Maxwell), cultured in floating net cages. Aquaculture, 15 : 273 - 287.

Vivekanandan, E., (1977). Effects of pO_2 on swimming activity and food utilization in *Ophioccphalus striatus*. J. Fish. Biol., 8 : 321 - 330.

Watanabe, W.O., Clark, J.H., Dunham, J.B., Wicklund, R.I. and Olla, B.L., (1990). Culture of Florida red tilapia in marine cages: the effect of stocking density and dietary protein on growth. Aquaculture, 90: 123-134.

Wahab, M.A., Rahman, M.M and Milstein, A. (2002). The effect of common carp, *Cyprinus carpio* (L.) and mrigal, *Cirrhinus mrigala* (Hamilton) as bottom feeders in major Indian carp polycultures. Aquaculture Res., 33, 547-556.

Warren, C. E. and Davis, G. E., (1967). Laboratory studies on the feeding, bioenergetics and growth of fish. In: The biological basis of freshwater fish production (S. D. Gerking, ed.), Blackwell Scientific Publications, Oxford. 175-214.

Zar, J. H. (1984). Biostatistical analysis, London : Prentice - Hall.

CHAPTER 17

A Review on the Heavy Metal Toxicity on the Freshwater Fishes

M. Aruchami, A.A. Sivakumar and R. Thangam

P.O. & Research Department of Zoology

Kongunadu Arts and Science College Coimbatore - 641 029, Tamil Nadu, India.

ABSTRACT

The impact of the heavy metals on aquatic organisms is due to the movement of pollutants from various diffuse or point sources which gives rise to coincidental mixtures in the ecosystem (Anderson and D' Apollonia, 1978), thus, posing a great threat to aquatic fauna especially to fishes, which constitute one of the major sources of protein rich food for mankind. Acute and sub lethal toxic effects of heavy metals have been investigated on the fishes and their impact on aquatic ecosystem have been studied by several authors.

Key words: *Review, heavy metal toxicity, freshwater fishes.*

INTRODUCTION

Aquatic environment gets contaminated with a variety of pollutants generated from diverse sources (industries, agricultural and domestic). Amongst the pollutants, pesticides, heavy metals and detergents are the major cause of concern for aquatic environment because of their toxicity, persistency and tendency to accumulate in organisms

Heavy metals are a group of 19 elements, which have many similar physical and chemical properties and are remarkably varying from the remaining 97 known elements. Among the 19 heavy metals Lead, Cadmium and Mercury do not have any biological significance or beneficial use and are known to be extremely toxic. Other toxic metals are Chromium, Copper, Manganese, Nickel, Tin and Zinc. Once dispersed in the biosphere, these metals cannot be recovered or degraded. Hence environmental effects of metal pollution are said to be permanent.

TOXICITY OF HEAVY METALS

Mercury

Banerjee (1985) studied the effects of heavy metal poisoning on peripheral Haemogram in *Heteropneustes fossilis* (Bloch); due to exposure to Mercuric sulphate (24-h LC_{50}). The length/breadth ratio of erythrocyte and erythrocyte nucleus did not a depart significantly from normal values and there was no significant change in erythrocyte and nuclear surface area. There was insignificant decrease in Total Erythrocyte count and Haemglobin content but significant decrease in packed cell volume and mean cell volume. ESR showed significant increase along with MCHC but the increase in MCH was insignificant. TLC increased significantly along with large lymphocytes and basophils but small lymphocytes decreased significantly. THC increased insignificantly but CT decreased significantly.

Palanichamy and Bhaskaran (1995) studied physiological responses of the fish, *Channa straitus.* It was observed that feeding, growth, protein conversion efficiency, oxygen consumption, protein, glycogen and SDH activity of heavy metals treated *Channa striatus* decreased with increasing concentrations of Mercury, whereas the protease activity increased.

Dhande and Patil (1998) reported the effect of heavy metal pollutants on the endocrine kidney of the fish, *Channa punctatus* (Bloch). Head kidney showed increased number of interrenal cells upto 10 days of exposure, but later on the number appeared to be slightly decreased. In Mercuric chloride treated fish the internal cells exhibited shrinkage. Chromaffin cells were distorted and lost their identify after 30 days of exposure to Mercuric chloride. Their number was increased. The hyperplasia observed in the experimental fish indicated more release of cortisol and catecholamines during the heavy metal stress to cope up with the toxic environment.

Patil and Dhande (2001) studied the effect of Mercury on ovarian recrudescence in the fresh water fish, *Channa punctatus* (Bloch). The exposure resulted in a statistically significant reduction in gonoado-somatic (GST) index after 30 days of exposure. Ovary exhibited significant decrease in cholesterol suggesting the possibility that the heavy metal either block hormone actions of the complex enzyme systems responsible for vitellogenensis.

Cadmium

Vijayram *et al.* (1989) studied the Cadmium induced changes in the biochemistry of an air breathing fish, *Anabas testudineus.* On short term exposure (3 days) a decrease was observed in liver and muscle glycogen content, accompanied by an increase in the lactic acid levels. Reduction in protein content and liver weight too was recorded. During long term exposure (54 Days) a reverse trend was observed.

Ruparalia *et al.* (1992) studied the Cadmium accumulation and biochemical alterations in the liver of freshwater fish, *Sarotherodon mossmbica* (Peters). Level of Cadmium in the liver tissue was maximum after 3 weeks exposure. Marked inhibition was noticed in the acidity of acid phosphatase during the course of study. Highest Cadmium concentration produced marked depletion of glycogen in liver at varying exposure periods. A reverse relationship between alkaline phosphatase activity and glycogen content was observed.

Cyril Arun kumar *el al.* (1994) reported the uptake and persistence of Cadmium in tissues of the freshwater fish, *Cyprinus carpio.* Accumulation of the xenobiotic Cadmium exhibited positive correlation with concentration as well as exposure period. Persistence of Cadmium in tissues was suggested to depend on availability and persistent of the contaminant in the medium.

Palanichamy and Baskaran (1995) studied physiological responses of the fish, *Channa striatus.* It was observed that feeding, growth, protein conversion efficiency, oxygen consumption, protein, glycogen and SDH activating decreased with increasing concentrations of Cadmium whereas the protease activity increased.

Thiruvallvan *et al.* (1997) observed residual accumulation of Cadmium in gill, liver and kidney tissues of carp fingerlings *of Cyprinus carpio.* Maximum acumulation of Cadmium was observed in liver and kidney tissues respectively. Gills showed least accumulation for toxicant. The rate of accumulation was positively related with concentration of the toxicant and the period of exposure.

Sastry *et. al.* (1997) observed the chronic toxic effects of Cadmium on some enzymological and biochemical parameters in *Channa punctatus.* The activities of LDH, MDH, GDH, XO, GPT and GOT were increased but the activities of SDH and PDH was reduced in gills after 15 days of exposure while the activities of all these enzymes were elevated after 30 days.

Vineeta Shukla and Sastry (1998) studied the biochemical, enzymological and haematological alterations produced on exposure of *Channa punctatus* to LC_{50} (11-2 mg/1) for 96 hr of Cadmium (Cd^{2+}). The fish were hypoglycemic and hypolactemic. The pyruvate content of blood and liver decreased in acute exposure. Depletion was noted in the total protein and glycogen content of liver and muscle, and the level of lactate in liver, while the level of muscle lactate and pyruvate increased after 96 hr. Activities of hexokinase, glucose - 6- phosphatase and lactate dehydrogenase decreased in liver. The activities of hexokinase and lactate dehydrogenase in muscle and glutamate-oxaloacetatc transaminase and glutamate-pyruvate transaminase in serum, liver and muscles increased. The activities of Succinate and Malate dehydrogneases in liver and muscles decreased after acute exposure.

Rao and Manjula Sree Patnaik (1999) studied the accumulation of Cadmium in the catfish, *Mystus vittatus* (Bloch). The concentration pattern in the different organs for each metal revealed that metal in muscle were low while on other organs the levels varied. The concentration of heavy metal from the natural habitat was also analysed and there was some evidence that accumulation in different organs was influenced by the variations in pH and it was more in acidic pH.

Roshan Ara Ashram *et al.* (2003) studied the toxicity effects of Cadmium chloride on some biochemical parameters of *Symbaranchus bengalensis* and *Anabas testudineus.* Following Cadmium chloride treatment (20ppm) the plasma protein level decreased in both the fishes. When fishes were treated with 30 ppm concentration of Cadmium chloride there was decrease in the plasma cholesterol levels of both the fishes.

Lead

Palanichamy and Baskaran (1995) studied physiological responses of the fish, *Channa striatus.* It was observed that feeding, growth, protein conversion efficiency, oxygen consumption, protein, glycogen and SDH activity were decreased with increasing concentrations of Lead, whereas the protease activity increased.

Mary Chandravathy and Reddy (1996) studied the Lead nitrate exposure changes in carbohydrate metabolism of freshwater fish, *Anabas scandens*. The changes in the activity patterns of the glycogen phosphorylases were clearly reflected in the metabolite levels. The enzymes and metabolites studied have failed to recover during initial recovery period. However on 15th day near normal levels were observed.

Rao and Manjula Sree Patnaik (1999) studied the Lead accumulation in the catfish, *Mystus vittatus* (Bloch). The concentration pattern in the different organs for Lead revealed that metal in muscle were low while on other organs the levels varied. The concentration of heavy metal from the natural habitat was also analyzed and there was some evidence that accumulation in different organs influenced by the variations in pH and it was more in acidic pH.

Nussey *et al.* (2000) studied the bioaccumulation of Lead in the tissues of the moggel, *Labeo umbratus* from Witbank Dam, Mpumalanga. Bioaccumulation of Lead varied between the gills, liver, muscle and skin. The gills generally had the highest metal concentrations, due to their intimate contact with the environment and their importance as an effector of ionic and osmotic regulation. The liver, in its role as a storage and detoxification organ can also accumulate high levels of metals. Muscles and skin accumulated much less metal concentrations. These two organs must be included in biomonitoring programmes because they are consumed by the general public. Accumulation of the metals decreased with an increase in fish length. Therefore, the smaller the fish the higher the body load of metals, due to various bioaccumulation processes.

Copper

Geetha *et al.* (1990) reported the toxic effects of heavy metal Copper on the fish, *Lepidocephalichthys thermalis*. In short term exposures more Copper in the tissue with the increased in concentration and exposure time was observed. Protein content of the fish could be taken as an indicator of Copper level.

Sastry *et al.* (1997) studied the chronic toxic effects of Copper on some enzymological and biochemical parameters in *Channapunctatus*. After 15 and 30 days in liver the activities of LDH, MDH, GDH, XO, GPT and GOT were increased but the activities of SDH and PDH was decreased. In muscle and kidney, the activities of all the enzymes were increased after 15 and 30 days of exposure to Copper. Inhibition in the activities of SDH, PDH and GDH was recorded in gills after 15 days of exposure while the activities of all these enzymes were elevated after 30 days.

Dhande and Patil (1998) studied the effect of heavy metal pollutants on the endocrine kidney of the fish, *Channa punctatus* (Bloch). Head kidney showed increased number of interrenal cells up to 10 days of exposure, but later on the number appeared to be slightly decreased. In $CuCl_2$ treated fish the interrenal cells exhibitied shrinkage. Chromaffin cells were distorted and lost their identity after 30 days of exposure to $CuCl_2$. Their number was increased. The hyperplasia observed in the experimental fish, indicated more release of cortisol and catecholamine during the heavy metal stress to cope up with the toxic environment.

Patil and Dhande (2001) studied the effect of Copper on ovarion recrudescence in the fresh water fish, *Channa punctatus* (Bloch). The exposure resulted in statiscally significant reduction in gonado-somatic index (GSI) differ 30 days of exposure. Extensive vacuolation in the oocartex, necrosis of oolemma and hypertrophy of follicular cells, degeneration of vitellogenic and hypertropy of follicular cells, degeneration of vitellogenic and large vitellogenic oocytes were observed. Further in fish the ovary exhibited significant decrease in cholesterol suggesting possibility that the heavy metal either block hormone actions or the complex enzyme systems responsible for vitellognensis.

Chromium

Sankar and Konar (1995) studied the effects of Chromium on fish and fish food organisms. It was observed that fish at sublethal concentration were reduced significantly; growth and reproduction offish were hampered by sublethal level.

Nussey *et. al.* (2000) studied the bioaccumulatioin of Chromium in the tissues of the moggel, *Labeo umbratus* from Witbank Dam, Mpumalanga. Bioaccumulation of Chromium varied between the gills, liver, muscle and skin. The gills generally had the highest metal concentrations, due to their intimate contact with the environment and their importance as an effecter of ionic and osmotic regulation. The liver, in its role as a storage and detoxification organ, can also accumulate high levels of metals. Muscles and skin accumulated much less metal concentrations. Accumulation of the metal decreased with an increase in fish length.

Manganese

Nussey *et. al.* (2000) studied the accumulation of Manganese in the tissues of the moggel, *Labeo umbratus (Cyprinidae)* from Witbank Dam, Mpumalanga. Bioaccumulation of Manganese decreased with increase in length of the fish.

Nickel

Nussey *et. al.* (2000) studied the bioaccumulation of Nickel in the tissues of the moggel, *Labeo umbratus (Cyprinidae),* from Witbank dam, Mpumalanga. Bioaccumulation of Nickel decreased with increase in length of the fish.

Iron

Arun *et. al.* (2000) observed the reduction in biochemical parameters (protein, carbohydrate, lipid) of the fish, *Oreochromis mossambicus* when exposed to Iron for 15 days. The detoxification mechanism in *Oreochromis mossambicus* was scrutinised by montioring the induction of xenobiotic conjugating enzyme (Glutathione-s-trnasferase) at every 5 days interval.

Zinc

Arun *et al.* (2000) observed the reduction in protein, carbohydrate and lipid during exposure of Zinc *Oreochromis mossambicus.* Mori *et. al.* (1990) studied the influence of water hardness and heavy metal (Zinc) on food utlization in *Cyprinus carpio* var. Communis.

It was noted that consumption, absorption, conversion and metabolism of *Cyprinus carpio* reared at 0.4 mg concentration of Zinc with 160 and 768 mg $CaCO_3$/l decreased, whereas at 384 and 576 $CaCO_3$/l were increased. The efficiency of absorption and conversion also followed the same trend.

Rao and Manjula Sree Patnaik (1999) studied the concentration pattern in the different organs for Zinc revealed that metal in muscle were low while on other organs the levels varied.

DISCUSSION

Fishes are exposed to elevated metal levels in an aquatic environment; they can absorb the bioavailable metals directly from the environment via, the gills and skin or through the injection of contaminated water and food. Metals in the fish are then transported by the bloodstream, which brings it into contact with the various organs and tissues (Vander Putte and Part, 1982). Fish can regulate metal concentration to a certain extent; whereafter bioaccumulation will occur (Health, 1991). Therefore, the ability of each tissue to either regulate or accumulate metals can be directly related to the total amount accumulated in that specific tissue. Furthermore, physiological differences and the position of each tissue in the fish can also influence the bioaccumulation of a particular metal (Kortze, 1997).

Mercury has insignificant decrease in total erythrocyte count and Haemoglobin content but significant decrease in packed cell volume and mean cell volume (Banariec, 1985). The increase in the concentration of Mercury leads to decrease in the oxygen consumption, protein, glycogen and SDH activity, excepting the protease which has increased activity (Palanichamy and Bhaskaran, 1995). Mercuric chloride affects the internal cells of the kidneys of the fishes. Initially the cells shrink and after continuous exposure chromaffin cells can be destorted as observed by Dhande and Patil (1998). The gonado-somatic index and the cholesterol content in the ovaries are also decreased.

Cadmium decreases the glylogen content in the muscle and liver and protein in the muscle (Vijaram *et al.*, 1989). Cadmium inhibits the acid phosphatase activity when present in the liver (Ruparalia *et al.*, 1992). Cadmium behaves similar to the Mercury when present in the tissues of the fishes by showing decrease in feeding, growth, oxygen consumption, protein, glycogen and SDH activity. Cadmium increases LDH, MDH, GDH, XO, GPT and GOT at the same time decreases the activities of SDH and PDH (Sastry *et al.*, 1997).

Lead behaves similar to that of Mercury and Cadmium by showing reduction in the feeding, growth, protein conversion efficiency, excepting the protease activity which increases (Palanichamy and Bhaskaran, 1995). Gills of fishes accumulate more amount of Lead than liver, muscle and skin, due to their intimate contact with the environment and their importance as an effector of ionic and osmotic regulation. The liver can also accumulate more levels of heavy metals as a storage and detoxification organ (Nussey *et. al.*, 2000).

Copper increases the activities of LDH, MDH, GDH, OX, GPT and GOT where as decreases the SDH and PDH activities as observed by Sastry *et. al.* (1997). Copper induces hyperplasia in the kidney cells of the fishes and releases more of cortisol and catecholamine due to this metal stress (Dhande and Patil, 1998).

Chromium affects the growth and reproduction of the fishes even in the sub-lethal level as observed by Sankar and Konar (1995).

Manganese and Nickel behave similar to the Chromium when present in the tissues of fishes. Higher amounts of Manganese and Nickel are found to be accumulated in gills and liver (Nussey *et al.*, 2000).

Iron and Zinc decrease the protein, carbohydrate and lipid content of the muscle and gill of fishes (Arun *et al*, 2000).

There are clear indications that heavy metals in the environment have increased to harmful levels. Exposure to these heavy metals show advêrse biogical conditions. Occupational exposure to heavy metals causes dermatitis, perforation of the nasal septum, and inflammation of the larynx and live in human beings. Hence biomonitoring is necessary for safe environment of the future.

REFERENCES

Arun, S., Thirumurugan, R., Visakan, R., Balamuragan, S., Arivazhagan, R., Subramanian, P., Prince Vijeya Singh and Malarvizhi, T. 2000. Toxicity induced biochemical modulations and phase II xenobiotic conjugating enzyme (GST) in *Oreochromis mossambicus*. Asian. J. S. Microbial. Biotech, and Environ. Sci. 2(34): 326-330.

Banerjee, V. 1985. Effect of heavy metal poisoning on peripheral haemogram in *Heteropneustes fossilis* (Bloch) : I. Mercuric sulphate (LC_{50}). Comp. Physiol. Ecol. 11 (4) pp. 173-176.

Cyril Arun Kumar, L., Vincent, S. and Ambrose, T. 1994. Uptake and persistence of the heavy metal Cadmium in tissues of the freshwater fish, *Cyprinus carpio*. Poll Res. 13(4): 361-364.

Dhande, R.R. and Patil, G.P. 1998. Effect of heavy metal pollutants on the endocrine kidney of the fish, *Channa punctatus* (Bloch). Proc. Acad. Environ. Biol. 7(2): 149-154.

Geetha, R., Kumaraguru, A.K. and Thatheyus, A.J. 1996. Toxic effects of heavy metal Copper on the fish, *Lepidocephalichthys thermalis* in short term exposures. Poll. Res.15(20): 151-153.

Health, A.G. 1991. Water Pollution and Fish Physiology. Lewis Publishers, Boca Raton, Florida, USA. 359pp.

Kortze, P. J. 1997. Aspects of water quality, metal contamintation of sediment and fish in the Olifants River, Mmpumalanga. M.Sc. Thesis, Rand Afr. Univ. South Africa.

Mary Chandravathy, V. and Reddy, S.L.N. 1996. Lead nitrate exposure changes in carbohydrate metabolism of freshwater fish. J. Environ. Biol. 7: 75 - 79.

Moni, D., Baskaran, P. and Palanichamy, S. 1990. Influence of water hardness and heavy metal (Zinc) on food utilization in *Cyprinus carpio*. Var. communis. J. Ecobiol. 2(1): 51-55.

Nussey, G., JHJ Van Vuren and Preez. 2000. Bioaccumulation of Chromium, Manganese, Nickel and Lead in the tissues of the moggel, *Lebeo umbratus* (Cyprinidae), from Witbank Dam, Mpumalanga. Water SA 26(2): 269-289.

Palanichamy, S. and Baskaran, P. 1995. Selected biochemical and physiological responses of the fish, *Channa punctatus* as biomonitor to assess heavy metal pollution in freshwater environment. J. Ecotoxicol. Environ. Monit. 5(2): 131-138.

Patil, G.P. and Dhande, R.R. 2001. Effect of Mercury and Copper on ovarian recrudescence in the freshwater fish, *Channa punctatus* (Bloch). J. Ecotoxicol. Environ. Monit. 11(2): 83-90.

Rao, L.M. and Manjula Sree Patnaik, R. 1999. Heavy metal accumulation in the catfish, *Mystus vittatus* (Bloch) from Mehadrigedda stream of Visakhapatnam, India. Poll. Res. 19(3): 325-329.

Roshan Ara Ashram, Manjula Yadav, Dillep Kumar and Yadav, B.N. 2003. Toxicity effects of Cadmium chloride on some biochemical parameters of *Symbaranchus bengalensis* and *Anabas testudineus.* Bio-Science Research Bulletin. 19(1): 75-78.

Ruparelia, S.G., Yogendra Verma, Meta, N.S. and Rawal, U.M. 1992. Cadmium accumulation and biochemical alternations in the liver of freshwater fish, *Saratherodon mossambica* (Peters). J. Ecotoxicol. Environ. Monit. 2(2): 129-136.

Sarker, U.K. and Konar, S.K. 1995. The combined effects of pesticide Thiodan, heavy metal Chromium and detergent Ekaline FI on fish and fish organisms. J. Environ. Biol. 19(1): 19-26.

Sastry, K.V., Sarita Sachdeva and Pratima Rathee. 1997. Chronic toxic effects Cadmium and Copper, and their combination on some enzymological and biochemical parameters in *Channapunctatus.* J. Environ. Biol. 18(3): 291-303.

Thiruvalluvan, M. Nagendran, N. and Charles Manoharan, A. 1997. Bioaccumulation of Cadmium and Methylparathion in *Cyprinus carpio* Var. communis (Linn.). Journal of Environment and Pollution. 4(3): pp.221-224.

Vander Putte, I. And Part, P. 1982. Oxygen and Chromium transfer in perfused gills of rainbow trout, *Salmo gairdneri* exposed to hexavalent Chromium at two different pH levels. Aquatic. Toxicol. 2: 31-45.

Vijayram, K., Geraldine, P., Varadarajan, T.S., George John and Loganathan, P. 1989. Cadmium induced changes in the biochemistry of an air breathing fish *Anabas testudineus.* J.Ecobiol. 1(4): 245-251.

Vineeta Shukla, and Sastry, K.V. 1998. Acute toxic effects of Cadmium on various parameters in the fresh water fish. Proc. Acad. Environ. Biol. 7(2): 155-160.

CHAPTER 18

MAINTENANCE AND BREEDING TECHNIQUES OF BLUE/THREE-SPOT GOURAMI, *TRICHOGASTER TRICHOPTERUS* (PALLAS, 1770)

Pradip Kumar, Y. C. Gupta and Vishal Tiwari*

Department of Zoology, B.S.A. College, Mathura-281001, Uttar Pradesh, India.
**Department of Zoology, R.B.S. College, Agra-282002, Uttar Pradesh, India.*

ABSTRACT

The present investigation deals with the study of maintenance and breeding procedures of a tropical freshwater, community aquarium fish, *Trichogaster trichopterus* (Blue/ Three-spot Gourami) of the labyrinth fish family. Their good community behaviour, ease of breeding, along with a peculiar nature of males to show parental care by building nest and guarding eggs until they hatch were some of the features noted during this tenure of work.

Keywords: *Fish, Trichogaster trichopterus (Blue/Three-spot Gourami), Maintenance and breeding techniques,*

INTRODUCTION

Trichogaster trichopterus, a native of South-East Asia belonging to the family Belontiidae and to the group of labyrinth fish is a robust, tropical freshwater, community aquarium fish. This is one of the easiest and most interesting fish to breed in an aquarium. Long conditioning periods are not necessary. Despite their exotic colourations, make care taking easy.

Most of the hobbyist identifies these fishes suprabranchial organ as a "Labyrinth", an apt designation as it shape implies. Aerial respiration is accomplished by gulping in atmospheric air, passing it back to the labyrinth organ, which is suffused with capillary beds. Owing to this, they breathe the oxygen directly from the atmosphere in the event the water they live in becomes depleted of oxygen or polluted. Another common name of these is

"Bubble Nest Builders", alluding to their mode of male nest building of floating bubbles and mucus, spawning and parental care.

MATERIALS AND METHODS

A set of 12 healthy adults male and female of 3 inches (7.5 cm) in the ratio 1:5 of *T. trichopterus* species of the same age group were put in a glass aquarium of 36 X 18 X 18 inches filled with potable water. The aquarium used was equipped with adequate illumination meeting the requirements of filter, an aerator pump and the accessories. The aquarium used was densely planted with live plants, such as hygrophila, limnophila, hydrilla on the sides and back of the tank, as well as the floating plants like riccia. Lots of hiding places for the females in the form of rocks and plenty of swimming area in the center to accommodate their lively antics was provided. As they are accomplished jumpers, a tight-fitting cover was used. An omnivore in nature, this fish accepts a wide range of foods, and therefore a wide range of foods, including small live foods, such as bladder worms, tubifex worms, earthworms, glass worms, brine shrimp as well as freeze-dried and flake foods was given. Supplement food with vegetable matter in the form of spirulina based foods or parboiled vegetables, such as romaine lettuce, zucchini or peas was also considered for them.

Another small sized aquaria of 24 X 12 X 12 inches was also prepared in the same manner possessing all the essential accessories as a tank, to separate the female into it once spawning is over. The optimum physico-chemical parameters of water were monitored periodically as they are considered to be the important factors influencing spawning. A water maintenance program of changing 20% of the water a week, should keep the fish in prime condition for breeding as well as displaying their best most vibrant colours. Anti-bacterial and anti-fungal medicines were also added in the aquarium water at some regular intervals.

RESULTS AND DISCUSSION

The morphological description of *T. trichopterus* (Blue/Three-spot Gourami) can be seen in Fig. 18.1. It was seen that they do well in water from 68 to 85 degrees, which makes them good fish for a natural tank, even one without extra filtration or aeration. The optimum pH was observed between 6 to 8.8 and hardness of 5 to 35 dGH. Some of the secondary sexual dimorphic features as observed in *T. trichopterus,* which were infact very promiscuous during mating period includes that the males were noted more colourful with longer and pointed dorsal fin, while the females were noted more rounded. The two long feelers of adult males were observed of bright orange colour, while that of females were noted of more yellow colour. The female had become slightly more rounded in the belly due to the formation of roe.

The male were introduced first. On the second day or the third day as the males had started building the nest, the females were introduced in the breeding tank. There was no need of any illumination, as they feel more comfortable without light. The building of "bubble nest" by the males is one of the best aspects of this fish. It was a huge effort, in which the nest consists of large pieces of vegetation, which were held together by many sticky bubbles, which the males 'spit' into the nest. The males was provided with sufficient amount of fresh vegetation, as only fresh leaves were used by them, which they rips from the plants by using a sawing action with their mouth.

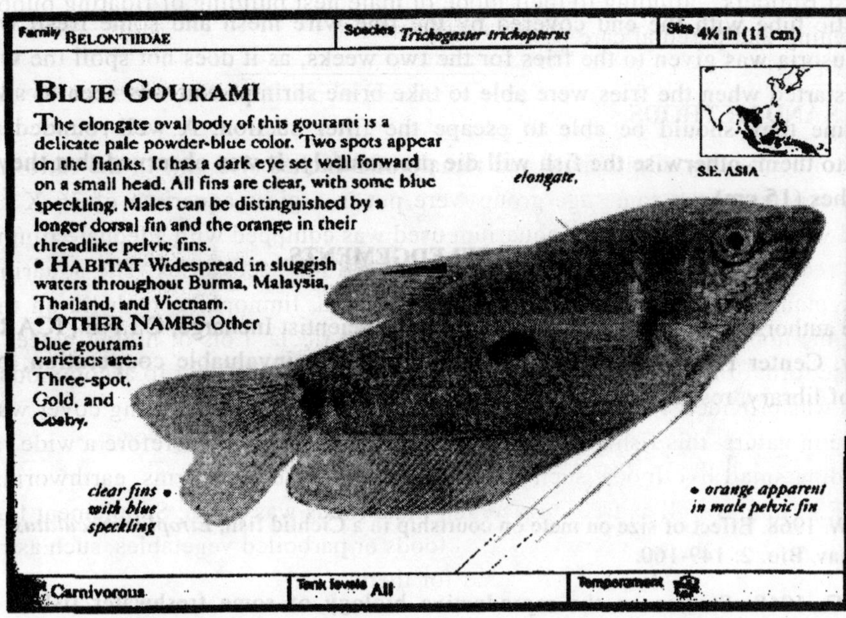

Fig. 18.1 Description of blue gourami, *Trichogaster trichopterus*.

The males and females regularly drift to the surface of the tank and releases an air bubble or swallow air. It was also observed that they used their feelers underneath their bodies to scope out new items or fish in the aquarium. As soon as the nest was built completely and enough bubbles had been amassed, the mating was started beneath the leaves. The male bends his body around the females, while she releases eggs. The fertilized eggs float up into the bubble nest. Once the eggs were laid down, the filter was shut off for a few days otherwise many fry get sucked into the filter. The males also immediately pick up the eggs and deposit them in the bubble nest. Again mating was observed and the eggs were again deposited. After all eggs had been released, the males start behaving violently towards the female and attack her. At this condition, the females were being transferred into the main tank.

It was observed that the eggs were guarded by the males, as for the two days they relentlessly pick up the eggs and blow them back into the nest and so on. It was also noted that during this period they were not eating the food normally, and therefore live worms were given to them, as they do not die so soon even if they were left uneaten.

Within a couple of days the larvae became apparent. The fry used their yolk sac for the first two days and during this period the male's looks after them by still picking them up from bottom and blowing them back into the nest. After that, the young were moved to a nursery tank and they were feed with cooked egg pressed through a piece of cheesecloth to make pieces small enough for them to eat. The nursery tank was kept covered for at least three to four weeks until "labyrinth organ" gets developed in them. It was done in order to avoid the surface of the tank to become chilled, which would be dangerous for the fries and they might get killed. The tank was kept between 75-78° F. Here water changes were not considered to be a must, as the fry do not produced much waste. However, part of the water was removed by using plastic tube with the end covered by the fine wire mesh and some fresh water was added. Infusoria was given to the fries for the two weeks, as it does not spoil the water. The filter was started when the fries were able to take brine shrimps and was seen to swim well. By this time they should be able to escape the filter suction. A well-rounded diet was important to them; otherwise the fish will die prematurely. It was observed that they reached upto 6 inches (15 cm).

ACKNOWLEDGEMENTS

The authors are thankful to Dr. A. K. Sharma, Scientist Incharge, C.I.F.E., ICAR Deemed University, Center Powarkheda, Hoshangabad (M.P) for invaluable cooperation, providing facilities of library, research literature during the present investigation.

REFERENCES

Barlow, G.W. 1968. Effect of size on mate on courtship in a Cichlid fish, *Etroplus maculatus*. Commun. Behav. Bio. 2: 149-160.

Bhatt, E. C. 1968. Studies on the reproductive biology of some freshwater fishes (part IV) *Heteropneustes fossilis* (Bloch). Ind. J. Fish. 15: 99-155.

David, A. 1965. Sexual dimorphism, fecundity and the food of the estuarine Bagrid, *Mystus gulio* (Ham.). Proc. Nat. Acad. Sci. Ind. 338: 385-410.

Gurchy, C.G. and Valadykov, V.D. 1968. Sexual dimorphism in anal fin of brown trout, *Salmo trutta* and close relatives. J. Fish. Res. Bd. Can. pp 813-815.

Heins, D.C. and Machado, M.D. 1993. Spawning season, clutch characteristics, sexual dimorphism and sex ratio in the redfin darter, *Ethestoma whipplei*. Amer. Midi. Nat. 129: 161-171.

Narejo, N.T.; Ali, S.S.; Jafri, S.I.H. and Hussain, S.M. 1998. Studies on the sexual dimorphism in Palla, *Tenualosa ilisha* from the river Indus. Pak. J. Zool. 30: 223-227.

Rosen, D.E. and Tucker, A. 1961. Evolution of a secondary sexual characters and sexual behaviour pattern in a family of viviparous fishes (Cyprinodontiformes: Poeciliidae). Copeia. 2: 201-212.

CHAPTER 19

Coastal Fisheries Activities of the Traditional Fishermen in the Contai Coast of Purba Medinipur District of West Bengal

Mrinmay Ghorai, Utpal Kumar Sar[1] and Bidhan C. Patra[2]

[1]*Department of Fisheries, Government of West Bengal, Meen Bhaban, Contai, Purba Medinipur*

[2]*Aquaculture Research Unit, Department of Zoology, Vidyasagar University, Midnapore - 721 102, West Bengal, India*

ABSTRACT

Studies on the coastal fisheries activities on the traditional fishermen in the Contai-I development Block of Purba Medinipur envisages ample opportunity of employment generation in all the 5 traditional landing centre (Khoties). In the present endeavours a detail observation was made on the crafts and gears, fish and fisheries and income level etc. certain suggestion have also been advocated for socio-economic improvement on the fisherfolk in this area.

Key Words : *Fisheries activities, traditional fishermen, landing centre, socio-economic improvement, fisherfolk.*

INTRODUCTION

The fishing profession has changed the oldest means of traditional livelihood for lakhs of people in India. The profession constitute an age old caste based activity among fishing community. The coastal belt of Purba Medinipur district of West Bengal is situated at the upper east coast of India in between the 21.45° N to 23.0°-N latitude and 87.3° E to 88.0°E longitude with a total area of 4151.64 sq. km. Nearly 70,000 traditional fishermen earn their livelihood through coastal fishing and subsidiary activities. The marine fish production is about 17.6% of the total fish production of West Bengal and 6.2% of the total marine production of India.

At present there are nearly 38 traditional fish landing centres recognised by Government of West-Bengal in five development blocks viz., Contai-I, Contai -II, Ramnagar -I, Ramnagar-II and Khajuri-II. Out of which Contai-I is one of the most important coastal fishing zone located almost middle of the coastal stretch of Purba Medinipur district.

There is no comprehensive information available on the fish and fisheries and socio-economic aspects of these landing centres except a few (CMFRI, 1987, DFWB, 1999 and Tewari *et. al.*, 2003).

With the objective to frame definite plans and programmes for socio-economic development of the community a detail observation on the fish and fisheries, crafts and gears in operation and some social aspects of the community has been carried out in five traditional landing centres under the Contai-I development block.

MATERIALS AND METHODS

In the present endeavours entire coast of Contai-I block was surveyed and the traditional landing centres locally known as khoties have been identified. To get the realistic information on fish and fisheries a set of questionnaires have interviewed with 100 numbers of fishermen of each landing centre like-time of fishing, inhabitance of the migrant people, number of fishermen engaged in actual fishing, number of craft and gears and their type, catch composition and catch per unit, seasonal occurrence and abundance of fishes if any, income level, family structure, educational level and so on.

Social structure, family set up, health hazards have also assessed with an another set of questionnaires. Profile of activities of fisher folk engaged in the khoties have been investigated with interrogation among the fishermen community, based on which clusterisation has been made.

RESULT AND DISCUSSION

It is a common observation in all the landing centre under the study area that all transient fishermen engaged in 'khoties' are coming from the nearby villages or adjoining blocks except Junput khoti where nearby 56% of the total migrant fisher folk came from Egra-II, Bhagwanpur-II and Khejuri -I Blocks etc. those are about 20-30 km. away from the 'khoti'. There are five registered landing centres in the 6.5 km. stretch of coastal belt under the Contai-I block viz. Junput, Haripur, Baguran Jalpai-I, Baguran Jalpai-II and Saula-I.

The Junput 'khoti' is the only traditional landing centre in the present study area which is connected with metalled road with bus route. Rest Khoties are partly or fully connected with non metalled road developed by Department of Fisheries Government of West Bengal. Generally operation of these khoties commence in the first week of October and lasts upto March end. The transient fishermen construct their temporary hut with 'hogla thatches' except Junput where few houses are permanent for storing dried fish and living. All 'khoties' are located on the sea beaches covering adjacent land under the holding of Fisheries department with land area ranging from 5 acres to 57 acres (Table -19.1).

Table 19.1

Showing the involvement of fishermen and catch in the study area under Contai-I block.

Name of the landing centre	Total land area use (Acres)	No. of fishermen engaged	No. of crafts use			No. of gears use	Total production during the year 2002-2003 (MT)
			Total	Mechnanised	Non mechanised		
1. Junput	57	4,847	176	109	67	746	7,105
2. Haripur	13	603	89	31	58	329	2,400
3. Baguram Jalpai-I	48	370	53	17	36	193	490
4. Baguram Jalpai-II	17	361	41	28	13	179	650
5. Salua-I	5	446	28	12	16	108	1,412

The biggest land area of 57 acres is being utilized by the Junput 'khoti' fisherfolk and the smallest having 5 acres of land under Saula-I 'khoti' with a total population of 4,847 and 466 respectively (Table -19.1). It is observed in the study area that per capita availability of land in the khoties are 0.011, 0.021, 0.129, 0.047 and 0.011 ac. in Junput, Haripur, Baguran Jalpai-I, Baguran Jalpai-II and Saula-I respectively. It is also recorded that per capita availability of marine fish as per production data are 1,465kg. , 3,980 kg., 1,324 kg., 1,800kg., and 3,165kg., in Junput, Haripur, Baguran Jalpai-I, Baguran Jalpai -II and Saula -I respectively (Table -19.1). The boats are mainly plank built 'dinghi' type (salti) with OAL 25 - 35 ft. may / may not be fitted with 2 cylinder engine generally killoskar (20 HP. -M-20) usually used for operation of stationary bag net- the 'Behundi' net. Fishing is done within 15km. from the coast, fixing the behundi net at a place for a fishing season in a row in the row at a suitable fishing ground generally nearby the landing centre. Number of boats in the different 'khoties' in operation is shown in the Table-19.1 envisages that maximum number i.e. 176 (109 mechanised and 67 non mechanised) and minimum is 28 (12 mechanised and 16 non mechanised) number of boats are in operation in Junput and Saula-I 'khoti' respectively.

Nearly 13 major important marine fishes were caught during the observation period 2002 - 2003 where only Bombay duck *(Harpodan nehereits)* constituted 17% of the total fish production (Table -19.2)

Table -19.2

Showing catch composition of different khoties in the study area during the year 2002-2003

Sl. No.	Name of the fish	Local name of the variety	Total catch during the year 2002-03 (MT)	% of over total catch	Remark
1.	*Tenualosa ilisha*	Ilish	301.925	2.5	Peak fishing season Aug-Nov.
2.	*Johnius* sp.	Bhola	1147.315	9.5	Used as raw and edible fish
3.	*Pristis microdon*	Shawfish	241.315	2.0%	Used as dry fish
4.	*Rastrelliger sp.*	Macrail	483.08	4.0%	Used as dry fish
5.	*Trichiurus sp.*	Patia	1570.01	13%	Used as dry fish
6.	*Pampus sp.*	Pompfret	664.235	5.5%	Used as raw and edible fish
7.	*Sardinella sp.*	Sardin	301.925	2.5%	Used as dry fish
8.	*Himantura* sp.	Shankar	181.155	1.5%	Used as raw and edible fish
9.	*Coilia dussumieri*	Ruli	543.365	4.5%	Used as edible fish
10.	*Harpodan nehereus*	Babla	2053.09	17%	Peak fishing season Dec-Jan
11.	*Setipinna phasa*	Tapra	1328.47	11%	Used as raw & edible fish
12.	*Penaeid sp.*	Chingri	905.775	7.5%	Used as dry fish
13.	*Acets indicus*	Gogua (Small shrimps)	301.925	2.5%	Used as fish meal
14.	*Misc.*	Misc.	2053.09	17%	-
		Total	12,077.09 MT		

Next to Bombay duck, the ribon fish *(Trichiurus sp)* formed 13% production in the catch. The other important fishes like *Setipinna phasa.*, *Johnius sp*, *Penaeid* prawn, pomfrets *(Pampus sp)*, *Coilia dussumieri*, *Rastrelliger* sp and *Tenualosa* sp formed 11%, 9.5%, 7.5%, 5.5%, 4.5%, 4%, 2.5% catch respectively in the fishery. Miscellaneous small fishes with a number of species constituted a very good fishery i.e. 17% during the study period which is mainly used for production of fish meal (Table - 19.2). A total of 12,077.09 MT of total landings were recorded in all the 5 khoties during the year 2002 - 2003 fishing season from first week of October and lasts upto March end.

The clusterisation of fisher folk in each khoti has methodically done based on the activities of the fishermen. Four types offish workers have recorded in all the five khoties. The active fishermen locally known as 'laya' are directly involved in catching fish from the fishing ground, where as, 'saudagars' those are involved in fish drying with the help of 'sramik' (labour).

Table -19.3

Showing particulars of fishermen population engaged in different khoties in the study area

Name of landing Centre	No. of fish workers involved in different fishing activities				% of literacy		% of income level (Rs.)			Health hazards
	Laya	Aratdar	Saudagar	Sramic	Male	Female	<10,000	>10,000 upto 25,000	>25,000	
Junput	176	50	307	4,314	31%	20%	72	18	10	Asthma, desenty, diarrhoea etc.
Haripur	89	11	21	522	45%	31%	82	12	06	
Baguran Jalpai-I	53	06	-	311	20%	15%	76	15	09	
Baguran Jalpai-II	41	-	25	295	25%	14%	79	11	10	
Salua-I	28	-	10	408	37%	21%	75	14	13	

In every khoti 'Aratdars' purchase dry fish and store for exporting to different indigenous market as well as abroad except Baguran Jalpai-II & Saula-I (Table -19.3) Maximum number of 'Sramik' were involved in Junput during the year 2002 - 2003 and nearly 69% of the sramik are women. In Baguran Jalpai - I, only 'layas' have recorded to take up the fish drying activity and the separate category of saudagars were not existed.

Sex wise literacy percentage has also recorded where in Baguran Jalpai - II shows the poorest performance i.e. 25% and 14% literate in male and female respectively. Highest percentage of literacy has been recorded in the Haripur khoti being 45% and 31% in male and female respectively. The income level group has also assessed through interrogation with huge number of fishermen and they are categorised in three groups like income level of Rs. < 10000, 10000 and upto Rs. 25000 and above 25,000 per annum. It is recorded that highest income group people are generally 'aratdar' and Junput shows the highest number under this category (Table-19.3)

From the foregoing observation it has been clear that per capital availability of land and the availability of fish is insignificantly related to different landing centers. It is reflected (Table-19.1) that per capita production in different landing centres (1465kg. 3980kg. 1324kg, 1800kg, and 3165 kg in Junput, Haripur, Baguran Jalpai-I, Baguran Jalpai-II and Saula-I khoties respectively) are influenced by the vicinity of the fishing ground and efficiency of the crafts and gears in operation for fishing. Though availability of fish per fisherman in Junput khoti is less (i.e. 1465kg./fisherman) in comparison to other khoties but availability offish per boat is much higher (40.36MT) in comparison (Table-19.1). Since Junput khoti which is directly connected with motor vehicle communication, the transient fishermen tends to migrate to this khoti even for involving themselves in subsidiary activities like fish drying. Regarding catch composition, *Harpodan nehereus* dominated in the catch during the year 2002 - 2003 with a production of 2053.09 MT out of total production of 12077.9MT. Most of this production is being dried in scaffoldings and transported mainly to north eastern hill region for human consumption. A bulk of production which is mostly unfit for human consumption is being dried on sandy beach for production of fish meal and used as ingredient of poultry feed. Among the khoties higher income group level (i.e. >Rs. 25000) fishermen has been observed in Saula-I khoti being 13%. It is also important that less number of fisherman were engaged in saula-I khoti and fishermen themselves trade dry fish thus, share the 'aratdars' margin which indicate traders earn a good amount even with a small capital invest. This observation emphasises that by mobilising. Co-operative society they can be further strengthen to sale their products through co-operatives. Regarding percentage of literacy, remote and isolated places without proper schooling facilities like 'Baguran Jalpai-I' and 'Baguran Jalpai-II' khoties shows lesser percentage of literacy i.e. 17.5% and 19.5% respectively in comparison to Haripur (38%), Soula-I (29%) and Junput (25.5%). From the above discussion we can conclude that though these traditional landing centres provide employment opportunity to a very good number of weaker section people still there is ample scope to develop socio-economic condition of these people with the following points.

(i) All 'khoties' are to be well connected with all-weather road to transport their product at a cheaper cost with easy.

(ii) Link school for six months may be setup atleast to literate fisher folk.

(iii) More emphasis has to be given to improve hygienic condition of all khoti particularly by constructing community latrine, sinking tube-well for drinking water.

(iv) Formation of co-operative society to strengthen fishermen to take cumulative decision and to improve their bargaining position.

(v) Values addition in fisheries product will not only increase their income but also enhance employment opportunity.

(vi) Improvement of the existing drying method may further improve the quality of the edible dry fish and thus may get better price in the market.

REFERENCES

CMFRI-1987, An appraisal of the marine fisheries of West Bengal Special Publication: No. 31 : 1-32.

Fishery Department Government of West Bengal, Millenium - 2000, Looking back, a Special Booklet Government of West Bengal: 1-25.

Tewary A., Bandyopadhyay, P. Sar, U.K. and Patra, B.C., 2003. Diversity of marine fin fish resources in the Midnapore Coast of West Bengal. In : Proceedings of National Seminar on "Marine Bio diversity as a source of food and Medicine". SDMRI Research Publication No. 3: 142-148.

CHAPTER 20

Effect of Distillery Effluent on Tissue Phosphatases Activity of Freshwater Teleost *Hypophthalmichthys Molitrix* (Val.) and *Catla Catla* (Ham.)

***C. Sarasu, *S. Andal**

**Dept. of Zoology, J.K.K. Natarajah College of Arts Science, Komarapalayam Namakkal dt. 638 183.*

M. Selvanyagam

Dept. of Zoology, Loyola College, Chennai-34.

ABSTRACT

Acid and alkaline phosphatase enzyme activity in different tissues of freshwater teleost may get altered under the stress of distillery effluent. The fingerlings of *Hypophalmichthys molitrix* and *Catla calta* used as experimental materials to study the impact of distillery effluent on the selected tissues like gill, muscle, liver, brain and intestine for an experimental period of 10, 20 and 30 days. Differential response of these two experimental species with regard to phosphatase activity was obvious and it could be used as pollution indicators in freshwater environment.

Key words: *Hypophthalmichthys molitrix (H.molitrix), Catla calta (C. catla), Acid phosphatase (AcP) and Alkaline phosphatase (AlP)*

INTRODUCTION

Several studies on the toxicity of industrial effluents and their constituents on fishes have been attempted in recent years. Though, distillery industries play an important role in social economy, their effluents have been found quite abnoxious, create and cause severe damage to the flora and fauna of water bodies (Panday *et al,* 1999)

Informations available on the toxicity of distillery effluent to freshwater teleost from Rahman (1972), Haniffa and Porchelvi (1985), Haniffa and Balakrishnan (1988), Haniffa and Augustin (1989) and Dhanapackiam and Manohar (1999). Very little work has been done on the impact of distillery effluent on enzyme activity of freshwater leleosts. Hence, an attempt has been made to study the effect of sublethal concentration of distillery effluent on phosphatases activity of freshwater teleosts *Hypophthalmichthys molitrix* and *Catla catla.*

MATERIALS AND METHODS

Live fringerlings of fresh water teleost *Hypophthalmichthys molitrix* and *Catla calta* (8-10 gms) were procured from Sirago Aqua Agri Farm, Nerunjipettai, near Bhavani and acclimatized to the ambient laboratory conditions. Distillery effluent was collected afresh from a local distillery near Bhavani and brought to the laboratory in Polythene containers and stored at 4°C.

Acute toxicity tests were carried out and suhlethal concentrations were derived (Finney 1971). After the experimental period of 10, 20 and 30 days the activity of Acid phosphatase (AcP) and Alkaline phosphatase (AlP) were assayed against the control in selected tissues like gill, muscle, liver, brain and intestine. The results were interpreted and discussed in the light of published literature.

RESULTS AND DISCUSSION

Acid phosphatase (AcP) and Alkaline phosphatase (AlP) activity in gill, muscle, liver, brain and intestine tissues of both experimental fishes *Hypophthalmichthys molitrix* and *Catpa catla* were recorded (Table 20.1 and 20.2) among which the highest AcP activity was observed in intestine and minimum in brain tissue. The findings from the present study were in good agreement with Verma *et al,* (1979), Haniffa and Porchelvi(1985)and Sharma (1990).

Table -20.1

Acid Phosphatase activity (KA units) in tissues of *Hypophthalmichthys molitrix* and *Catla catla* exposed to sublethal concentration of distillery effluent

Period of Exposure (days)	Tissues	Hypophthalmichthys molitrix		Catla catla	
		Control	Experimental	Control	Experimental
1	2	3	4	5	6
	Gill	17.421±0.001	16.200±0.138	17.427±0.001	16.479±0.011
10	Muscle	9.650±0.002	8.570±0.124	9.651±0.002	8.700±0.679
	Liver	18.667±0.001	17.160±0.052	18.690±0.073	17.420±0.019

1	2	3	4	5	6
	Brain	8 928 ± 0.000	8.222 ± 0.045	8.930 ± 0.094	8.270 ± 0.022
	Intestine	22.196 ±0.005	15.815 ±0.001	22.233 ± 0.001	16.200 ± 0.047
	Gill	17.427 ± 0.001	15.200 ± 0.036	17.440 ± 0.097	15.604 ±0.051
20	Muscle	9.657 ± 0.002	8.153 ± 0.100	9.664 ± 0.008	8.320 ± 0.038
	Liver	18.702 ±0.001	16.500 ± 0.048	18.702 ±0.001	16.740 ±0.043
	Brain	8.940 ± 0.007	7.153 ±0.068	8.940 ± 0.243	7.450 ±0.016
	Intestine	22.233 ± 0.001	14.300 ± 0.037	22.240 ±0.010	15.352 ± 0.037
	Gill	17.460 ± 0.057	13.450 ± 0.123	17.458 ± 0.050	13.850 ±0.025
30	Muscle	9.680 ± 0.001	7.460 ± 0.057	9.700 ± 0.052	7.800 ± 0.040
	Liver	18.553 ± 0.094	15.100 ± 0.085	18.720 ± 0.053	15.350 ±0.023
	Brain	8 960 ± 0.056	6.280 ± 0.064	8.952 ± 0.048	6 750 ± 0.036
	Intestine	22.260 ±0.015	13.550 ± 0.048	22.270 ± 0.025	13.900 ±0.037

Values are the mean ± S.E.

Table - 20.2

Alkaline Phosphatase activity (KA units) in tissues of *Hypophthalmichthys molitrix* and *Catla catla* exposed to sublethal concentration of distillery effluent

Period of Exposure (days)	Tissues	Hypophthalmichthys molitrix		Catla catla	
		Control	Experimental	Control	Experimental
1	2	3	4	5	6
	Gill	22.817±0.211	20.500 ± 0.030	23.000 ± 0.126	21.180 ±0.067
10	Muscle	12.396 ± 0.001	11. 260 ±0.040	12.601 ± 0.060	11.500 ±0.065
	Liver	35.265 ± 0.001	33.000 ± 0.049	35.751 ±0.063	33.570 ± 0.048
	Brain	11.180 ±0.016	10.400 ± 0.062	11.300 ±0.033	10.600 ± 0.041
	Intestine	44.320 ± 0.029	40.250 ± 0.034	44.600 ±0.016	41.300 ±0.019
	Gill	23.481 ±0.000	19.000 ±0.019	23.097 ±0.081	20.403 ±0.015
20	Muscle	12.420 ± 0.024	10.120 ±0.032	12.700 ± 0.044	10.900 ±0.015
	Liver	35.269 ± 0.000	32.150± 0.048	35.800 ± 0.628	32.610 ± 0.046

1	2	3	4	5	6
	Brain	11.196 ±0.001	10.150 ±0.038	11.352 ±0.037	10.450 ± 0.045
	Intestine	44.340 ± 0.015	38.000 ± 0.060	44.750 ± 0.022	38.650 ± 0.043
	Gill	23.495 ± 0.000	18.350 ± 0.062	23.150 ±0.062	19.320 ±0.021
30	Muscle	12.450 ± 0.037	9.080 ± 0.020	12.752 ± 0.042	9.750 ± 0.047
	Liver	35.300 ± 0.035	30.950 ± 0.052	35.900 ± 0.017	32.000 ±0.051
	Brain	11.250 ±0.044	9.220 ± 0.024	11.400 ±0.053	9.602 ± 0.042
	Intestine	44.350 ± 0.044	37.180 ±0.038	44.902 ± 0.996	37.850 ± 0.067

Values are the mean ± S.E.

Acid phosphatase is a hydrolytic enzyme concerned with the process of trans phosphorylation and also helps in autolysis of the cell after its death (Novikoff 1961). The most probable reason for the decrease in the activity of AcP could be uncoupling of oxidative phosphorylation. It has been explained on the basis of chemical (Pressman, 1963) and chemostatic (Mitchel, 1961) interactions. According to Pressman (1963), uncouplers promote the conductivity of protons within mitochondrial mechanisms and subsequently prevent a gradient across the membrane. Permeability, disruption of normal functioning of cell organelle like lysosome and mitochondria and different represser mechanisms associated with toxicity resulted in significant change in the level of enzyme AcP in fish tissues examined.

Alkaline phosphatase a brush border enzyme responsible for the phosphorylation, more likely by hydrolysis of sugar phosphate esters in the diet at alkaline p^H and mediate membrane transport. The declining trend of AcP in the present investigation are in consistent with observations of Shobarani *et al.*, (2001) which showed that the inhibition of ALP activity with consecutive impairment of phosphorylation. Uncoupling might have also occurred which promotes the spilling of an energy rich intermediate compound prior to ATP production (Ramesh *et al*, 1994).

According to Sastry and Sharma (1980) A1P activity in the brain is involved in the permeability process in the fish and they form a part of the enzyme system taking part in the active synthesis of protein from the nucleic acid complex. The present findings are in accordance with Haniffa and Porchelvi (1985) suggesting that inorganic and organic salts in the distillery effluent might have interfered with respiration by coagulation of gill mucous, asphyxiation and inhibition of enzyme system at mitochondrial level.

Many polluters like distillery effluent at low dosages affect the metabolison of biota by altering normal enzyme activity to varying degrees and hence enzymes are considered to be logical candidates to be used as bioindicators. The present findings indicated differential response of the different species of freshwater teleosts to distillery effluent stress and *H.molitrix* is more susceptible than *C.catla.*

ACKNOWLEDGEMENTS

The help rendered by the management and Principal, J.K.K. Natarajah College of Arts & Science and the financial assistance by U.G.C. are gratefully acknowledged.

REFERENCES

Dhanapackiam, P. and L. Manohar, 1999 : Realtive toxicity of Tannerey and distillery effluents to the fish *Labeo rohita* (Ham.) as determined by 96 hr. LC_{50} and behavioral analysis. Convergence 1 (3 & 4) : 103-112.

Finey, D.J., 1971 : Probit analysis. Cambridge University Press, London, pp.333.

Haniffa, M.A. and M.Porchelvi, 1985: Effects of distillery effluents on *H.Fossilis* (Bloch): and *S.monambicus* (Peters). Proc. Indian Acad. Sci. (Anim Sci.) 5 (2) : 155-161.

Haniffa, M.A. and Maria Augustin, 1989 : Oxygen consumption, surfacing frequency and distance travel in *Rana malabahca* tadpoles exposed to distillery effluent. J.Environ., Biol. 10 (2) : 139-147.

Haniffa, M A. and K.Balakrishnan, 1988 : Bioenergetics of young common carp *C.carpio* exposed to distillery effluent. Poll. Research. (7): 103-104.

Novikoff, A.B., 1961: The cell Eds. J.Brachet and A.P. Misskey Academic Press. New York.

Pandey, G.C., S.Nezaliya and B.K. Dwivedi, 1999: Phybotoxic effects of distillery effluent on the *Allium cepa.* I.J. Environ & Ecophah 2 (3): 265-267.

Ramesh, M., R.Manavalaramanujam and K.Sivakumari, 1994 : Effect of Vegetable oil factory effluent on alkaline phosphatase activity in a fresh water teleost fish *Cyprinus carpio (*var) communis. I.J. Environ. Hlth. 36 (3): 192-196.

Rahman, A.A. 1984 : Effect of distillery effluent on the biota of rivery Cauvery. Problems and solutions Education., T.N.Veziroglu. Environmental Science Publ., B.V.Amstadam.

Sastry, K.Y, and S.K. Sharma, 1980 : Diazinon effect on the activites of brain enzymes from *Ophiocephalus (C.Punctatus)*. Bull. Environ. Contam. Toxicol. 24 : 326-332.

Sharma R.M., 1990 : Effect of endosulfan on acid and alkaline phosphatase activity in liver, kidney and muscles of *C.gachua.* Bull. Environ. Contam. Toxicol 44 : 443-448.

Shobarani, A., R. Sudharsan, T.N.Reddy, P.U.M. Reddy and T.N. Raju, 2001: Activity levels of phosphatase of the fish *Tilapia monambica* subjected to arsenite toxicity. Poll. Respondents. 20(1): 9-12.

Verma, S.R., A.K.Tyagi and R.C.Dalela, 1979 : Effects of distillery waste in some freshwater teleosts: Biochemical studies, Environ., Pollut. 19 : 225-228.

CHAPTER 21

STUDY ON THE SEXUAL DIMORPHISM OF *POECILIA RETICULATA* AND *GAMBUSIA AFFINIS* (POECILIDAE)

Vishal Tiwari
Department of Zoology, R.B.S. College, Agra-282002 (Uttar Pradesh), India.

ABSTRACT

The present investigation aimed to study the universal phenomenon of sexual dimorphism among *Poecilia reticulata* (Peters, 1860) and *Gambusia affinis* (Baird & Girard, 1853) belonging to the family Poeciliidae. It facilitates mutual recognition for mating purposes as mating preferences may be determined by male courtship display, body size, colouration, pigmentation patterns, and visual cues. The presence of developed and functional gonopodium in adult males and gravid spot in adult pregnant females of both these fishes also facilitates their distinction and recognition for mating.

Keywords : *Pisces, Poeciliidae, P. reticulata, G. affinis, sexual dimorphism.*

INTRODUCTION

Sex recognition along with courtship and pair formation are essential parts of vertebrate reproduction. The important stimuli involved in the sex recognition are considered as visual, olfactory, auditory and tactile stimuli. Visual recognition may result from sexual dimorphism or from differences in the behaviour of males and females within a species. Secondary accessory sexual dimorphic features had always played a key role in facilitating mutual distinction and recognition among organisms including fish.

In fish, the relations of the sexes are very promiscuous during mating period. The characteristics of sexual dimorphism, which enable identification of the either sexes, may be categorized as primary and secondary. Primary sexual characters include testes and their ducts in the male; and ovaries and their ducts in the female. Secondary sexual characters are of two kinds: those that have no primary relationship within the reproductive act at all, and those that are definitely assessory to spawning. The secondary sexual characteristics and

accessory structures, which are used in courtship, clasping or intromission, develop simultaneously with the re-adding of the primary sexual products. With this view point *P. reticulata* and *G. affinis,* an important aquarium, viviparous fishes had been undertaken to investigate their sexual dimorphic features which are still inadequate and needs due emphasis in terms of mates.

The important works on the said field of studies are from Rosen and Tucker (1961); David (1965); Barlow (1968); Heins and Machado (1993); and that of Narejo *et al.* (1998).

MATERIALS AND METHODS

A set of 30 healthy adult males and females of *P. reticulata* and *G. affinis* of the same age group in the ratio 2:1 respectively were put in two separate glass aquariums of 24X12X12 inches filled with potable water. These were equipped with adequate illumination, meeting the requirements of plants, an aerator pump and the accessories. Water quality parameters were monitored periodically. Aquarium water of both was changed at regular intervals. A readymade food along with some live food (dried daphnia etc.) was given to the fishes. Anti-bacterial and anti-fungal medicines were also added in the aquarium water at some regular interval.

RESULTS AND DISCUSSION

The observations on the sexual dimorphic features of male and female of *P. reticulata* and G. *affinis* revealed following results (Table 21.1).

Table 21.1

Sexual dimorphic features of *P. reticulata* and *G affinis.*

Characters	*P. reticulata*		*G. affinis*	
	Male	Female	Male	Female
1	2	3	4	5
Body form, shap, size	Normal 3 cm (1.2 inches)	Pot-Bellied 5-6 cm (2 to 2.4 inches)	Normal 3-4 cm (1.2 to 1.6 inches)	Pot-Bellied 5-7 cm (2 to 2.8 inches)
Colouration	Brighter and brilliantly colourd, having rainbow colours	Drab coloured	Brighter translucent grey; dusted with black spots and blotches; bluish sheen on the flanks, belly silvery	Grey coloured; dull olivaceous pigments
Dorsal fin	Vastly extended; more colourful	Not extended; Normally coloured	Colourful (orange or red)	Have black spots.

1	2	3	4	5
Caudal fin	Brightly coloured elongated with gross variations in shapes	Short, fan-shaped; not much coloured	Gross variations in shapes brightly coloured	Short, fan-shaped; grey
Pelvic fin	Designed to help swing of the gonopodium, functional	Normal	Designed to help swing of the gonopodium, orange	Normal
Gonopodium	Developed, rod like	Girthy, large, normal shaped	Functional, developed rod like	Girthy, large, normal shaped
Gravid spot	Absent	Present in pregnant females	Absent	Present in pregnant females

Body Form, Shape and Size

It acts as one of the important secondary sex-determining characteristic, which had always played a key role in facilitating mutual distinction and recognition. The males were noted normal, while the females were noted much more pot-bellied, with round and tight stomach, which in late pregnancy becomes cuboidal in shape. Males of *P. reticulata* were recorded 3 cm (1.2 inches), while females 5-6 cm (2 to 2.4 inches). Males of *G. affinis* were 3 to 4 cm (1.2 to 1.6 inches), while females 5 to 7 cm (2 to 2.8 inches). Thus the males of both the species were noted considerably smaller than the females. Females need to be large because she literally carries the precious cargo of dozen or more babies.

Body Colour

Males of *P. reticulata* were noted brighter and brilliantly coloured. The females were noted drab coloured, and they do not share the male's rainbow colours. Depending upon the particular strain of *P. reticulata,* it may be red, blue, green, black or any combination of these and any other colours too. This fish is known as the "rainbow fish" or "peacock fish" because of the numerous colour patterns on their body.

G. affinis was noted as modestly coloured fish. The body colour observed was mainly grey, dusted with black spots and blotches. The males were noted brighter, translucent grey, with a bluish sheen on the flanks, black olive-brown, belly silvery. Females were similarly grey coloured but exhibit dull olivaceous pigments. Females were also being recognized with light orange pelvic fins or a yellow-orange spot at vent.

The bright colour of males of both these species may be used for advertising their health, virility and presence to females.

Dorsal, Caudal and Pelvic Fin

In *P. reticulata,* dorsal fins of males were noted vastly extended and more colourful than females. Caudal fin of males is brightly coloured, elongated with gross variations in shapes, while that of females is short, fan-shaped and not so much coloured. The variations in shapes and colours of it in males are supposed to catch the attention of females towards them. Pelvic fin of male fish was noticed to designed in such a way that they aid in stabilizing and directing the swing of the gonopo dium, while that of a female is normal.

In *G. affinis,* the dorsal fin of male was more colourful (orange to dark red). A female has black spots on the dorsal fin. Caudal fin of males also exhibits variations in shapes for the same purpose. Pelvic fin of females was normal, while that of males is designed to aid in stabilizing and directing the swing of the gonopodium.

Gonopodium (modified anal fin)

It is the most significant external discriminatory feature between males and females of both these species, separating livebearers from all other types of fish. It acts as a valuable index in finding relationship between the groups of fishes. In order to fertilize the female, the anal fin of both the males of *P. reticulata* and *G. affinis* has developed into a functional structure to insert the packets of sperms into the female fish. It is called "gonopodium". The anal fin of females of both these was noted girthy, large and normal shaped. Gonopodium was examined supported internally by bony structures and moved backward, forward and sideways.

Gravid spot

A pregnant female of both the species can be easily identified by a characteristic black spot or gravid spot, present at the posterior end of the gut near to the urinogenital vent. It acts as a valuable index in discriminating males and females of these species, as it is absent in case of males. It may be considered equivalent to the womb, but unlike mammals, the eggs contain an embryo well furnished with nutritive elements on which it feeds during development. The eyes of the developing fry were visible through the thin walls of the gravid spot.

ACKNOWLEDGEMENTS

The author is thankful to Dr. A. G. Ponniah, Director, National Bureau of Fish Genetic Resource (N.B.F.G.R.), Lucknow and Dr. A. K. Sharma, Scientist Incharge, C.I.F.E., ICAR Deemed University, Center Powarkheda, Hoshangabad (M.P) for invaluable cooperation, providing facilities of library, research literature during the present investigation.

REFERENCES

Barlow, G.W. 1968. Effect of size on mate on courtship in a cichlid fish, *Etroplus maculatus*. *Commun. Behav.* Bio. 2: 149-160.

David, A. 1965. Sexual dimorphism, fecundity and the food of the estuarine Bagrid, *Mystus gulio* (Ham.). *Proc. Nat. Acad. Sci. Ind.* 338: 385-410.

Gurchy, C.G. and Valadykov, V.D. 1968. Sexual dimorphism in anal fin of brown trout, *Salmo trutta* and close relatives. *J. Fish. Res. Bd. Can.* pp 813-815.

Heins, D.C. and Machado, M.D. 1993. Spawning season, clutch characteristics, sexual dimorphism and sex ratio in the redfin darter, *Ethestoma whipplei. Amer. Midi. Nat.* 129: 161-171.

Narejo, N.T.; Ali, S.S.; Jafri, S.I.H. and Hussain, S.M. 1998. Studies on the sexual dimorphism in Palla, *Tenualosa ilisha* from the river Indus. *Pak. J. Zool.* 30: 223-227.

Rosen, D.E. and Tucker, A. 1961. Evolution of a secondary sexual characters and sexual behaviour pattern in a family of viviparous fishes (Cyprinodontiformes: Poeciliidae). *Copeia. 2:* 201-212.

CHAPTER 22

Malathion and Temperature Induced Alterations on Some Blood Parameters of Freshwater Gobiid Fish, *Glossogobius giuris* (Ham.)

G.V. Venkataraman*, P.N. Sandhya Rani ** and P.S. Murthy **

** Department of Bio-Technology, V.V.Pura College of Science, Bangalore - 560 004, India.*

*** Department of Zoology, Bangalore University, Bangalore - 560 056, Karnataka, India.*

ABSTRACT

Environmental stress has been known to interfere with physiological functions such as growth, development, reproduction and immune system in fish. The immune system is greatly related to the water quality and external environmental factors. Though the effect of pesticide on various tissues of freshwater fishes is well known, our knowledge on temperature and pesticide combination effect on haematological parameters of fishes is not fully understood. In the present study, the blood of *Glossogobius giuris* was examined after maintained in different temperature (20, 27 and 30°C) and also treated with sub lethal concentrations of malathion (0.05, 0.25 and 0.5 ppm) for short term (24 to 96 h) duration. Exposure of *G.giuris* to sublethal levels of an Organophosphorus pesticide, malathion at 27 °C (Laboratory conditions) produced a time and dose dependent decrease in RBC count, haemoglobin content, and mean corpuscular volume, leading to microcytic anaemia and an increase in WBCs. During 24 and 48 h exposure with malathion at 20 °C, RBCs, Hb, Ht was found to be decreased while WBCs, MCV and MCHC increased. However in 72 and 96 h at 20 °C showed an inverse trend. At 30°C significant decreasing tendency in RBCs, WBCs, Hb content, Ht and MCV was noticed in the treated fish with 0.5 ppm of malathion and a slight variations of colour index and MCHC concentration. The differential count of leucocytes showed a significant increase in the population of neutrophils, monocytes and macrophages at higher concentration (0.5 ppm) for longer duration (96 h) at 30 °C the significant results are discussed.

Key words : *G.giuris,* malathion, temperature, RBC, WBC, Ht, Hb, exposure.

INTRODUCTION

The extensive use of Pesticides, Insecticides, Herbicides/Seedicides and Fungicides are being promoted by Government of India to enhance the crop production to meet the demand of the growing population. The rationale behind this is the eradication of undesirable insects, pests, weeds and herbs for increased yield but on the other hand, these lead to large scale mortality of one of the most important aquatic fauna, the fishes forming stable food to the mankind. All these inturn cause severe ecological imbalance. These chemicals reach aquatic systems by direct application, spray drift from ground or aerial, atmospheric fallout, run-off from agricultural land and discharge of effluents. The concentration of pesticides can produce chronic histopathological changes in aquatic organisms. Survival and dispersal of organisms therefore depend upon their successful adaptation to the changed environmental conditions in aquatic media Mount and Stephen (1967) and Venkataramana *et al.*, (2001).

These pesticides persist in the environment and are found to be toxic to various aquatic animals including the fishes (Kumar, 1994). The physiological changes occurring as a result of pesticide intoxication may be due to alterations in pH, temperature, dissolved oxygen and hardness of water. Temperature variations influence the degree of toxicity of various pesticides in aquatic organisms. (Pascoe *et al*, 1986 and Nemcosk *et al.*, 1987). Haematological parameters have been used for detecting diseases, dietary deficiencies and environmental stress (Srivastava, 1989). Such study provides valuable information in the health management of fishes. Hence in the present investigation an attempt has been made to study the haematological changes in response to malathion - temperature combination.

MATERIALS AND METHODS

The freshwater gobiid fish *G. giuris* (body length 8-10 cms body wt. 20-25 gm) were collected from local kelavarapalli dam near Bangalore by using cost and gill nets (mesh : 10 mm) and acclimated to in lab conditions in glass aquaria (60" x 30" x 20") for 15 days. Fish were fed with standard laboratory diet during acclimatization and toxicity test. Commercial grade of malathion [0,0-dimethyl, S-(1,2- dicarboethoxy ethyl) phodophorodithioate] was used in the study. Acetone : Ethanol (1:1 v/v) was used as a solvent for the preparation of stock solution, which was further diluted to required concentrations of 0.05, 0.25 and 0.5 ppm in water was made by adapting the dilution techniques as outlined in APHA AWWA WPCF (1985).

The acclimatized fishes were divided into four experimental groups. The first three groups of fishes were placed in freshwater having sublethal concentrations (0.05, 0.25 and 0.5 ppm) of malathion and the fourth group in freshwater with no pesticide served as control. The second set was also divided into four groups and the first three groups were subjected to different concentrations of malathion (0.05, 0.25 and 0.5 ppm) with temperature variation, fourth group served as control (20°C and 30°C). The acclimated fishes were starved for 24h prior to their use in the experiment and were not fed during the course of experiments (Dalela *et al.*, 1978). The tap water was changed on alternate days and the concentration of pesticide was maintained. In the experimental and control fishes, the blood was collected by severing the caudal peduncle in the intervals of 24, 48, 72 and 96h, causing minimum stress to the fish, and the blood was drawn with the help of 1 ml graduated syringe fitted with 2.4 G x 1" needle coated with sodium citrate as anticoagulant.

Total counts of erythrocytes (TEC), leucocytes (TLC), haemoglobin content (Hb), haematocrite (Ht) and blood indices like, mean corpuscular volume (MCV), mean corpuscular haemoglobin (MCH), mean corpuscular haemoglobin concentration (MCHC) and colour index (Cl) were determined following the methods of, Blaxhall and Daisley (1973) and Wintrobe (1977). From the total count of erythrocyte and leucocyte, the erythrocyte and leucocyte ratio (RBC / WBC ratio) was calculated.

RESULTS

At 27 °C (lab conditions) the decrease in RBC count and Hb concentration was recorded in 0.5 ppm of malathion exposure when compared to control. Similarly a decrease of PCV and MCV was also observed in higher concentration of malathion (0.5 ppm Table 22.1). On the other hand, in 0.5 ppm the WBC count had increased from $20.58 \pm 0.13 \times 10^3$ cmm to $39.89 \pm 0.38 \times 10^3$ cmm for 24 to 96 h respectively (Table 22.1).

After 24 and 48 h of exposure of malathion (0.05 and 0.5 ppm) at 20 °C the RBC count significantly decreased, and was more conspicuous in higher concentration (0.5 ppm) when compared to control (RBC : $2.85 \pm 0.04 \times 10^6$ cmm, $2.51 \pm 0.60 \times 10^6$ cmm for 24 and 48 h). However at 20 °C with high concentration (0.5 ppm), RBC count had increased during 72 and 96 h (RBC: $2.40 \pm 0.04 \times 10^6$ cmm and $2.45 \pm 0.08 \times 10^6$ cmm) when compared to control (RBC : $2.22 \pm 0.06 \times 10^6$ cmm and $1.93 \pm 0.13 \times 10^6$cmm) (Table 22.2).

At 30 °C with malathion combination a decrease in RBC count was noticed in 0.05 ppm of malathion (RBC : $2.07 \pm 0.03 \times 10^6$ cmm, $1.52 \pm 0.05 \times 10^6$ cmm, 1.50 ± 10^6 cmm, $1.46 \pm 0.40 \times 10^6$ cmm) for 24, 48, 72 and 96 h respectively. The significant decrease was also observed in 0.5 ppm malathion (RBC : $1.66 \pm 0.06 \times 10^6$ cmm for 24h, $1.43 \pm 0.10 \times 10^6$ cmm for 48 h, $1.36 \pm 0.08 \times 10^6$ cmm for 72 hand $1.37 \pm 0.09 \times 10^6$ cmm for 96 h) when compared to control (RBC : $2.32 \pm 0.04 \times 10^6$, $1.78 \pm 0.05 \times 10^6$, $1.77 \pm 0.07 \times 10^6$ cmm, $1.69 \pm 0.09 \times 10^6$ cmm) for 24, 48, 72 and 96 h respectively (Table. 22.3).

The haemoglobin content (Hb) and PCV (Ht) was found to be decreased after exposure to 0.05 ppm of malathion and further decreased was also noticed with higher concentration (0.5 ppm) at 20 °C (Hb : 6.52 ± 0.19 g per cent and 6.16 ± 0.72 g per cent). On the other hand an increase of Hb content in 0.5 ppm malathion (Hb: 7.88 ± 0.17 g per cent and 9.20 ± 0.84 per cent) and PCV at 20 °C was observed (PCV: 6.86 ± 0.15 per cent and 8.00 ± 0.51 per cent) for 72 and 96 h. (Table 22.2).

At 30° C the Hb concentration had decreased from 24 to 96 (Hb: 6.88 ± 0.19 g per cent, 6.24 ± 0.17 g per cent, 5.96 ± 0.48 g percent and 5.44 ± 0.78 g per cent) for 24, 48, 72 and 96h respectively after exposing to 0.05 ppm malathion. The maximum decrease in Hb content was observed at 72 and 96 h of treatment (5.16 ± 1.28 g per cent and 3.92 ± 1.23 g per cent) with 0.5 ppm of malathion. The PCV values were found to have significantly declined in the malathion (0.05 ppm) at 30 °C, after exposure for 24 to 96 h (PCV : 6.16 ± 1.37 per cent, 4.76 ± 2.00 per cent, 4.00 ± 0.94 per cent and 3.88 ± 0.99 per cent) than the controls. The decrease in the values were more conspicuous in 0.5 ppm (PCV : 5.04 ± 0.97 per cent to 3.32 ± 0.91 per cent) for 24 to 96 h (Table 22.3).

Table 22.1

Effect of sublethal concentrations of malathion on haematological parameters of *Glossogobius giuris*. (mean ± S.E.)

Parameters	RBC ($X10^6$/cmm)	WBC ($X10^3$/cmm)	RBC/WBC ratio	Hb(g%)	Ht(%)	CI	MCV (μm^3)	MCH (pg)	MCHC(%)
24 Hours									
Control	3.33±0.07	14.08±0.45	23.65±0.38	11.21±0.08	10.0±0.28	0.200±0.01	25.06±0.40	6.01±0.28	24.00±1.22
0.05	2.44±0.02	16.24±0.56	15.02±0.91	10.53±0.04	7.93±0.08	0.196±0.02	23.06±0.28	5.91±0.10	25.41±0.94
0.25	2.23±0.03	16.96±0.67	13.14±0.31	10.02±0.07	7.39±0.35	0.186±0.02	15.58±0.95	5.58±0.17	35.82±1.21
0.5	2.04±0.03	20.58±0.13	10.12±0.48	9.40±0.07	4.26±0.04	0.243±0.01	18.20±0.42	7.30±0.11	40.14±1.32
48 Hours									
Control	3.11±0.19	15.08±0.35	20.62±0.68	10.18±0.09	9.43±3.5	0.220±0.09	26.08±0.48	6.21±0.38	25.55±2.11
0.05	2.76±0.06	18.88±0.22	14.61±1.00	10.42±0.04	7.39±0.13	0.189±0.12	22.53±0.85	5.67±0.03	25.20±1.00
0.25	2.28±0.05	27.41±0.22	8.31±1.31	9.86±0.07	4.47±0.07	0.200±0.02	16.19±0.41	6.01±0.36	31.13±2.27
0.5	2.07±0.04	29.33±0.34	7.05±0.97	9.06±0.08	4.06±0.03	0.262±0.02	19.61±0.25	7.87±0.08	40.14±1.31
72 Hours									
Control	3.20±0.29	15.43±0.41	20.73±0.40	9.89±0.06	9.38±3.00	0.241±0.02	25.40±0.40	6.91±0.27	25.71±2.66
0.05	2.45±0.06	27.76±0.36	9.81±1.12	9.48±0.02	6.71±0.07	0.186±0.02	21.78±0.58	5.59±0.07	25.67±0.39
0.25	2.08±0.90	34.45±0.34	6.03±0.71	8.53±0.01	3.90±0.02	0.212±0.01	15.91±0.73	6.44±0.07	40.51±2.06
0.5	1.98±0.05	36.58±0.35	5.41±0.41	8.17±0.05	3.80±0.12	0.256±0.02	19.69±0.47	7.7±10.12	40.15±0.01
96 Hours									
Control	3.00±0.46	14.48±0.68	20.71±0.65	10.41±0.09	10.00±1.2	0.250±0.02	24.48±0.48	6.48±0.33	26.78±2.06
0.05	2.38±0.05	30.93±0.05	7.69±0.79	7.70±0.02	5.82±0.08	0.173±0.02	20.93±0.66	5.23±0.19	25.01±0.41
0.25	2.24±0.09	39 50±0.65	5.67±0.51	8.50±0.04	3.69±0.05	0.189±002	16.47±052	5 69±0.24	34.57±1.52
0.5	1.65±0.07	39.89±0.38	4.23±0.38	7.06±0.03	2.58±0.18	0.226±0.01	15.63±0.29	6.42±0.33	41.08±0.98

Table 22.2

Effect of sublethal concentrations of malathion on haematological parameters of *Glossogobius giuris.* at 20°C (mean ± S.E.)

Parameters	RBC ($X10^6$/cmm)	WBC ($X10^3$/cmm)	RBC/WBC ratio	Hb(g%)	Ht(%)	CI	MCV (µm³)	MCH (pg)	MCHC(%)
24 Hours									
Control	2.85±0.04	22.40±0.88	12.72±1.83	8.52±0.90	6.16±0.09	0.996±0.07	21.61±1.78	29.89±0.81	138.31±0.31
0.05	2.16±0.22	17.90±1.64	12.06±2.40	8.24±0.25	4.20±0.99	1.271±0.09	19.44±2.11	38.14±0.91	196.10±0.41
0.25	1.72±1.50	27.50±4.1	6.25±1.71	8.00±0.80	4.40±0.75	1.425±0.08	25.58±2.61	46.51±1.20	181.81±0.91
0.5	1.56±0.60	29.40±1.93	5.30±2.79	6.52±0.19	4.68±0.22	1.392±0.02	30.00±1.99	41.79±0.41	139.31±0.81
48 Hours									
Control	2.51±0.60	25.28±1.60	9.92±1.70	8.00±0.06	5.28±0.83	1.062±0.07	21.03±2.41	31.87±0.91	151.51±1.11
0.05	1.83±0.06	23.72±1.26	7.71±2.00	7.60±0.34	4.08±0.76	1.384±0.05	22.29±3.11	41.53±0.68	186.27±1.33
0.25	1.53±0.02	30.00±2.00	5.10±3.22	6.52±0.20	4.00±0.40	1.420±0.03	26.14±3.67	42.61±0.81	163.00±0.17
0.5	1.34±0.06	30.85±0.60	4.34±4.20	6.16±0.72	4.28±0.23	1.532±0.06	31.94±2.66	45.97±1.11	143.92±1.11
72 Hours									
Control	2.22±0.06	22.23±0.57	9.98±2.82	7.56±0.73	5.20±0.77	1.198±0.09	23.42±1.77	35.94±0.71	153.46±0.91
0.05	2.38±0.32	27.84±1.80	8.54±3.33	7.98±0.20	6.40±0.18	1.058±0.04	26.89±2.77	31.76±0.86	118.12±1.60
0.25	2.24±0.25	22.72±0.41	9.85±3.63	6.94±1.36	5.80±0.26	1.032±0.07	25.89±3.11	30.98±1.12	119.65±1.32
0.5	2.40±0.04	18.06±0.83	13.28±2.46	7.88±0.17	6.86±0.15	1.063±0.04	27.77±3.00	31.90±1.25	114.86±2.00
96 Hours									
Control	1.93±0.13	20.64±0.88	9.35±3.31	6.36±0.73	5.94±0.71	1.098±0.02	30.77±1.97	32.95±1.33	117.07±0.71
0.05	2.15±0.07	22.72±1.36	9.46±2.10	8.28±0.31	7.28±0.33	1.283±0.04	33.86±1.66	38.51±0.95	113.73±1.21
0.25	2.20±0.05	18.88±0.76	11.65±3.70	7.96±0.38	6.96±0.44	1.205±0.05	31.63±1.81	36.18±0.71	112.93±1.61
0.5	2.45±0.08	16.80±1.78	14.58±2.80	9.20±0.84	8.00±0.51	1.251±0.06	32.65±2.22	37.55±1.91	115.00±1.51

Table 22.3

Effect of sublethal concentrations of malathion on haematological parameters of *Glossogobius giuris* at 30°C (mean ± S.E.)

Parameters	RBC ($X10^6/m^3$)	WBC ($X10^3/m^3$)	RBC/WBC ratio	Hb(g%)	Ht(%)	CI	MCV (μm^3)	MCH (pg)	MCHC(%)
24 Hours									
Control	2.32±0.04	15.92±0.48	14.57±1.34	8.04±0.17	8.56±1.78	1.150±0.02	36.8912.16	34.65±0.45	93.92±0.39
0.05	2.07±0.03	24.08±0.72	8.59±2.08	6.88±0.19	6.16±1.37	1.107±0.07	29.75±2.27	33.23±0.67	111.68±0.86
0.25	1.82±0.02	24.44±1.42	7.44±1.27	6.40±1.15	5.84±1.49	1.170±0.09	32.08±3.05	35.16±0.29	109.58±0.32
0.5	1.66±0.06	17.12±1.72	9.69±2.79	5.92±0.23	5.04±0.97	1.188±0.07	30.36±2.86	35.66±1.00	117.46±0.83
48 Hours									
Control	1.78±0.05	24.24±2.0	7.34±1.80	6.64±0.21	4.72±1.42	1.243±0.02	26.51±1.86	37.30±0.47	140.67±0.63
0.05	1.52±0.05	13.68±1.54	11.11±1.70	6.24±0.17	4.76±2.00	1.368±0.22	31.31±2.60	41.05±0.38	131.09±0.26
0.25	1.48±0.09	12.80±2.40	11.56±3.27	5.84±0.29	3.80±2.11	1.315±0.11	25.67±2.63	39.45±1.31	153.68±0.66
0.5	1.43±0.10	10.40±3.40	13.75±4.30	4.96±0.49	3.60±2.20	1.155±0.07	25.17±2.65	34.68±0.71	137.77±1.20
72 Hours									
Control	1.77±0.07	22.32±2.20	7.93±2.80	6.60±0.40	4.52±1.20	1.242±0.09	25.53±1.66	37.28±0.78	146.01±0.91
0.05	1.50±0.10	16.56±3.12	9.05±3.86	5.96±0.48	4.00±0.94	1.324±0.19	26.66±2.11	39.73±0.73	149.00±0.94
0.25	1.38±0.07	14.08±2.40	9.80±3.08	5.24±0.48	3.44±0.96	1.265±0.07	24.92±4.13	37.97±0.92	152.32±0.66
0.5	1.36±0.08	11.60±0.09	11.72±2.44	5.16±0.28	3.44±1.20	1.264±0.09	25.29±5.55	37.94±0.69	150.00±1.05
96 Hours									
Control	1.69±0.09	24.72±0.31	6.83±2.14	5.92±0.88	4.40±1.72	1.167±0.06	26.03±5.93	35.02±0.71	134.54±0.62
0.05	1.46±0.40	18.00±0.21	8.11±2.72	5.44±0.78	3.88±0.99	1.241±0.05	26.57±6.11	37.26±1.10	140.20±0.72
0.25	1.40±0.30	14.80±0.31	9.45±1.81	4.28±1.22	3.64±0.71	0.907±0.05	24.97±3.95	29.31±1.31	117.58±1.40
0.5	1.37±0.09	12.48±0.41	10.97±1.91	3.92±1.23	3.32±0.91	0.953±0.04	24.23±2.85	38.61±0.89	118.07±1.61

The fish exposed to malathion (0.5 ppm) at 20 °C for 24 and 48 h the WBC count was found to have increased (WBC : 29.40 ± 1.93 x 10^3 cmm and 30.85 ± 0.60 x 10^3 cmm) and decreased in the higher concentration (0.5 ppm) of malathion and at 20 °C for 72 and 96 h (WBC : 18.06 ± 0.83 x 10^3 cmm and 16.80 ± 1.78 x 10^3 cmm) than the control (Table 22.2). At 30 °C significant decrease of WBC count was observed in higher (0.5 ppm) concentration of malathion after 24 to 96 h of exposure (WBC : 17.12 ± 1.72 x 10^3 cmm to 12.48 ± 0.41 x 10^3 cmm) for 24, 48 72 and 96 h respectively (Table 22.3). There were variations in values of MCV, MCH and MCHC when the fish exposed to sublethal concentrations (0.05, 0.25 and 0.5 ppm) of malathion for 24 to 96 h at 20°C and 30°C (Table 22.2 and 22.3).

The differencial count of leucocytes showed a significant changes in the percentage of lymphocytes, neutrophils, monocytes and macrophages after treating the fish with sublethal concentrations of malathion at 20 °C, 27 °C and 30 °C illustrated in Table - 22.4, 22.5 and 22.6. The increase in these parameters was time and concentration dependent. The percentage of lymphocytes gradually decreased when fish were exposed for 24 to 96 h at 20 °C and 30 °C compare to control. The neutrophil population showed an inverse trend during 72 and 96 h and significant increase in monocyte and macrophage population at higher concentration (0.5ppm) for longer duration (96 h) (Table 22.6).

Table 22.4

Differential leucocyte count (%) of *Glossogobius giuris* after exposure to sublethal concentrations of malathion (mean ± S.E)

	Lymphocytes	Neutrophils	Monocytes	Macrophages
	1	2	3	4
24 Hours				
Control	78.52±1.00	18.82±0.41	2.64±0.26	-
0.05	*71.72*±1.65	23.79±0.72	1.72±0.15	2.75±0.39
0.25	72.75±1.03	22.06±0.77	2.41±0.13	2.75±0.22
0.5	65.14±1.68	26.97±0.28	3.73±0.17	4.14±0.21
48 Hours				
Control	76.00±1.87	20.51±0.51	1.48±0.11	2.00±0.31
0.05	76.22 ±2.15	17.81±0.81	2.33±0.28	3.64±0.28
0.25	71.20±3.12	24.05±0.89	2.84±0.20	1.09±0.22
0.5	66.50±1.97	27.93±0.57	2.75±0.22	2.41±0.13

	1	2	3	4
72 Hours				
Control	76.32±1.40	20.92±0.51	1.59±0.18	1.10±0.27
0.05	74.62±3.10	22.78±1.18	1.41±0.15	1.17±0.22
0.25	72.56±1.36	23.58±2.92	2.05±0.22	1.79±0.13
0.5	71.93±2.65	24.52±1.69	2.22±0.13	2.22±0.13
96 Hours				
Control	72.45±1.65	22.97 ±0.88	1.71±0.11	2.85±0.27
0.05	76.38±1.10	20.00±0.62	1.94 ±0.28	1.06±0.11
0.25	68.02±1.95	28.17±1.03	2.53±0.25	1.26±0.12
0.5	70.58±2.66	25.55±0.40	2.38±0.14	1.47±0.14

Table 22.5

Differential leucocyte count (%) of *Glossogobius giuris* after exposure to sublethal concentrations of malathion at 20°C (mean ± S.E)

	Lymphocytes	Neutrophils	Monocytes	Macrophages
	1	2	3	4
24 Hours				
Control	87.06±.359	10.88±0.59	1.40±0.13	0.94±0.11
0.05	64.45±1.91	31.29±1.56	2.65±0.25	1.59±0.21
0.25	72.85±4.16	20.28±0.97	4.57±0.32	2.28±0.15
0.5	56.97±4.88	35.34±1.24	4.65±0.39	3.02±0.52
48 Hours				
Control	69.49±0.67	28.00±0.48	1.50±0.22	1.00±0.31
0.05	69.82±1.60	24.41±1.14	2.81±0.39	1.06±22
0.25	80.43±2.77	15.65±0.95	2.39±0.14	1.08±0.12
0.5	65.12±2.28	30.44±1.53	2.82±0.32	1.61±0.22

	1	2	3	4
72 Hours				
Control	78.93±3.23	17.58±0.48	2.04±0.17	1.43±0.13
0.05	65.84±2.09	31.22±1.15	1.52±0.39	1.40±0.21
0.25	75.12±3.75	20.04±1.69	2.89v0.28	1.94±0.22
0.5	63.56±1.83	32.44±1.71	1.86±0.21	2.12±0.32
96 Hours				
Control	71.22±0.93	22.65±1.44	2.38±0.28	2.04±0.21
0.05	74.16±4.40	18.62±2.50	3.48±0.40	3.0210.39
0.25	70.76±5.22	24.59±1.29	3.01±0.22	1.62±0.13
0.5	71.02±5.53	20.98±1.05	4.75±0.60	3.26±0.33

Table 22.6

Differential leucocyte count (%) of *Glossogobius giuris* after exposure to sublethal concentrations of malathion at 30°C (mean ± S.E)

	Lymphocytes 1	Neutrophils 2	Monocytes 3	Macrophages 4
24 Hours				
Control	91.28±2.30	6.42±0.45	1.14±0.12	1.14±0.15
0.05	73.06±5.72	23.34±1.55	1.95±0.41	1.63±0.35
0.25	70.59±2.96	24.34±1.48	3.55±0.28	1.68±0.30
0.5	70.39±1.68	23.44±0.28	3.31±0.17	2.97±0.27
48 Hours				
Control	68.14±1.50	25.00±0.63	3.47±0.58	3.46±0.35
0.05	73.73±3.77	23.41±1.57	1.42±0.32	1.42±0.32
0.25	70.69±2.51	23.99±0.97	2.74±0.39	2.56±0.44
0.5	67.36±2.05	22.83±0.77	5.46±0.31	4.34±0.22

	1	2	3	4
72 Hours				
Control	66.76±2.14	23.44±0.94	5.63±0.20	4.15±0.41
0.05	69.35±0.69	25.48±0.89	3.78±0.28	2.38±0.28
0.25	73.07±5.00	22.30±1.41	2.76±0.51	1.84±0.28
0.5	68.95±2.16	23.31±0.57	3.86±0.31	2.86±0.14
96 Hours				
Control	70.31±2.56	21.91±0.57	4.10±0.16	3.65±0.56
0.05	78.99±1.53	16.17±0.57	2.52±0.21	2.31±0.34
0.25	69.96±2.40	25.16±1.80	2.85±0.48	2.01±0.39
0.5	62.26±0.90	30.38±0.60	3.46±0.22	3.88±0.25

DISCUSSION

Blood analysis forms an useful tool for determining physiological and pathological conditions of the fishes. *G. giuris* exposed to sublethal concentrations of malathion at 27 °C (lab conditions) resulted in a significant decrease in RBCs count leading to anaemia as a result of inhibition of erythropoiesis, haemosynthesis and increase in the rate of erythrocyte destruction in haemopeoietic organs. Similar reports have been made by Goel and Kalpana (1985) and Sampath *et al.,* (1998) on *Heteropneustes fossilis* and *Oreochromis mossambicus* after exposure to zinc and copper. Srivastava and Shashikala (1979) have shown that exposure to sublethal concentration of Pb produced haemolytic anaemia leading to the lysis of erythrocytes, decrease in Hb content and PCV value in *Colisa fasciatus.* Natarajan (1981) reported a reduction in Hb content RBC count and PCV values, resulting in hypochronic anaemia due to deficiency of iron and decreased utilization for Hb synthesis. In the present study on *G.giuris* sublethal concentration of malathion might have impaired the synthesis of iron as a result destruction of mature RBC and reduction in RBC count causing macrocytic anaemia. The increased WBC count in treated fishes may be due to the inclusion of thrombocytes in the leucocyte population.

The anaemic condition recorded in the present study could be due to the destruction of mature RBC or inhibition of erythrocyte production. Such a decrease in RBC and anaemic response have been observed by Koundinya and Ramamurthy (1979) in *Sarotherodan mossambicus* and *Tilapia mossambicus* after exposure to lethal concentrations of sumithion. Chouhan *et al.,* (1983) in *Puntius ticto* treated with herbicide, Lal *et al.,* (1986) in *H. fossilis* exposed to malathion. It is evident that the reduction in number of RBC and Hb content and decrease in MCV values might have caused microcytic anaemia in *G. giuris.* Varadharaj *et al.,* (1993) have opined that a reduction in the number of RBC and Hb content with an increase in

MCV and MCHb values might cause macrocytic anaemia in *O. mossambicus*. Since the Hb and RBCs are oxygen carrying devices, the quantitative decrease in their levels might have led to the dearrangement of the oxidative metabolism with a concomitant decrease in the tissues of respiratory potential. An increase in the total WBC count and MCHC content in *G. giuris* after exposure to different sublethal concentrations of malathion coincides with the observations made by Radhakrishnan and Prasad (1994). In the present investigation MCV significantly decreased which coincide with the findings of Surendra Sigh and Bhati (1994).

The results in the present study indicate that the fishes exposed to malathion at 20 °C the total number of leucocytes was significantly increased and RBC count, Hb concentration and PCV gradually declined during 24 and 48 h. On the other hand RBC count and PCV increased gradually when the fishes were exposed to malathion with the increased period of exposure (72 and 96 h) at 20°C. However the total WBC, MCHC and MCH significantly decreased when compared to those of controls. The significant decrease in MCV with simultaneous increase in MCHC might have led to microcytosis in *G.giuris* as pointed out by Haws and Goodnight (1962) and Joshi *et al.,* (1980). The decrease in RBC count, Hb content and PCV in *G. giuris* subjected to temperature (20 °C) and malathion toxicity are in agreement with the observations made by Slicher and Pickford (1968) in *Fundulus heteroclitus* and Anthony (1961) in *Carassius auratus.*

The fishes exposed to malathion at 30 °C the haematological parameters except MCHC were found to have decreased with the increase in the exposure period. This phenomenon suggests that a high temperature and malathion toxicity may cause anaemic and macrocytic changes in G. *giuris* when compared to low temperature and malathion stress. Similar results have been recorded by Joshi *et al.,* (1980) in *C.batrachus.* In *G. giuris* a decrease in WBC count could have led to a fall in the resistance for malathion and temperature (at 30 °C) stress, and this suggests that, malathion in combination with high temperature has an adverse effect resulting in anemic and macrocytic conditions. An increase in malathion concentration and exposure period resulted in the increased leucocytosis. An increase of leucocytes might be due to the increase in the population of lymphocytes, neutrophils and basophils. The neutrophil increase in fish exposed to malathion may be due to tissue damage. Mahajan and Dheer (1979) and Sampath *et al.,* (1998) reported that neutrophils showed great sensitivity to changes in the environment and were the more important of leucocytes. The increase in lymphocytes and monocytes suggests that these fish becomes stimulated under pesticide stress to fight against the pollutants in the environment. However in the differential count gradual decline in the number of lymphocytes in fish exposed to malathion at 30 °C for longer duration (96 h). This significant find may be due to continuous stress. The present observations is in accordance with Mc Leay (1973) who suggested that stress in attributable to a decrease the number of circulating lymphocytes.

ACKNOWLEDGEMENTS

The authors are grateful to the Chairman, Department of Zoology, Bangalore University, Bangalore - 56 for the laboratory facilities and co-operation.

REFERENCES

Anthony, E.H 1961. The oxygen capacity of gold fish. *Carassius auratus* in relation to the thermal environment. *J. Exp. Biol.*, 38 : 93-107.

APHA, AWWA and APCF. 1985. Standard methods for the examination of water and waste water, 16th Edition, American Public Health Association, Washington, D.C.

Blaxhall, P.C. and Daisley, K.W. 1973. Routine haematological methods for use with fish blood. *J. fish Bioi.*, 5: 771-782.

Chouhan, M.S., Verma, D. and Pandey, A.K. 1983. herbicide induced haematological changes and their recovery in a fresh water fish, Puntius ticto (Ham.), Comp. *Physiol. Ecol.*, 8(4):249 - 251.

Dalela, R.C., Bhatnagar, M.C., Tyagi, A.K.and Verma, S.R. 1978. Adenosine triphosphatase activity in few tissues of a freshwater teleost, *Channa gachua,* following in vivo exposure to endosulfan. *Toxicology,* 11: 361 - 368.

Goel, K.A. and Kalpana, G. 1985. Haematological characteristics of *Heteropneutes fossilis* under the stress of zinc. *Indian. J. Fish.*, 36:256-259.

Haws, T.G. and Goodnight, C.J. 1962. Some aspects of the haematology of two species of catfish in relation to their habitats. *Physiol Zool.*, 35: 8-17.

Joshi, B.D., Chaturvedi, L.D. and Dabral, R. 1980. Some haematologic values of *Clarias batrachus* following its sudden transfer to varying temperature. *Indian J. Exp. Bioi,* 18:76-77.

Koundinya, P.R. and Ramamurthy, R. 1979. Effect of Organophosphate sumithion (Fenthion) on some aspects of carbohydrate metabolism in a freshwater fishes *Sarotherodon, mossambicus* and *Tilapia mossambicus* (Peters).J. *Expt.*, 15: 1632-1633.

Kumar, A.V. 1994 Endosulfan induced biochemical and pathophysiological changes in freshwater fish, *Clarias batractus* (Linn.) Ph.D. Thesis, Osmania University, Hyderabad, India.

Lal, A.S.B., Anitakumari, S. and Sinha, R.N. 1986. Biochemical and haemotological changes following malathion treatment in the freshwater catfish *Heteropneustes fossilis* (Bloch). *Environ. Pollut.*, 42(A): 151-156.

Mahajan.C.L. and Dheer., J.M.S. 1979. Cell types in the peripheral blood of an airbreathing fish *Channa punctuatus.* J. fish Biol., 14: 481-487.

Mc.Leay., D.J. (1973) : Effect of 12 and 25 days exposure to kraft pulpmill effluent on the blood and tissue of Juvenile coho salmon *Oncorhynchus Kisutch.* J. Fish. Research Board. Can., 30: 395-400.

Mount, D.I. and Stephan, C.E. 1967. A method for establishing acceptable toxicant limits for fish malathion and the butoxy - ethanol ester of 2,4-D. *Trans. Amer. Fish.Soc.*, 21:185-193.

Natarajan, G.M. 1981 Changes in the biomodal gas exchange and some blood parameters in the air-breathing fish *Channa striatus* (Bloch) following lethal exposure to metasystox (Dimeton). *Curr. Sci.*, 50:40-41.

Nemcosk, T., Orban, L., Asztalos, B. and Vig, E, 1987. Accumulation of pesticides in the organs of carp *Cyphnus carpio L.* at 4 and 20°C *Bull. Environ. Contam. Toxicol.*, 39: 370-378.

Pascoe, D., Evans, S. and Woodworth, J. 1986. Heavy metal toxicity to fish and the influence of water hardness. *Arch. Environ. Contam. Toxicol.*, 15:481-485.

Radhakrishnan, S. and Prasad, G. 1994. A unique haematological response of *Oreochromis mssambicus* to sublethal concentrations of the insecticide, Ekaluk EC-25. Proceedings of the Indian National Science Academy Part B. Biological Sciences 60(6): 499-503.

Sampath, K., James, R. and Akbar Ali, K.M. 1998. Effects of copper and zinc on blood parameters and prediction of their recovery in *Oreochromis mossambicus. Indian. J. Fish.*, 45(2): 129-139.

Slicher, A.M., and Pickford.G.E., 1968. Temperature controlled stimulation of haemopoiesis in a hypophysectomized cyprinodont fish *Fundulus heteroclitus. Physiol.Zool.,41:293-297.*

Srivastava.A.K. and Shashikala, M. 1979. Blood dyscrasia in a teleost, *Colisa fasciatus* after acute exposure of sublethal concentration of lead. *J. fish boil.* 14:199-203.

Srivastava, A.K. 1989. A review of landmarks in teleostean haematology, Proc. Natl., Symp. Emerg. Tr. Anim. Haematol., 165-177.

Surendra Singh and Bhati, D.P.S. 1994. Effect of zinc chloride on certain morphological parameters of the blood in *Channa punctatus* (Bloch). Poll. Res., 13(14):381-384.

Varadharaj, G., M.A. Subramanian and R. Nagarajan 1993. The effect of sublethal concentration of paper and pulp mill effluent on the haematological parameters of *Oreochromis mossambicus* (Peterns). J. Environ., Biol., 14(4):321-325.

Venkataramana, G.V., D.U. Anandhi, and P.S.Murthy. 2001 : Effect of malathion on the haematology of gobild fish, *Glossogobius giuris* (Ham). Indian. J. Trends in Life Sciences. 16(1): 19-26.

Wintrobe, M.M. 1977. clinical haematology. Henry Kipton. London, 448.

CHAPTER 23

APPLICATION OF ITK IN FISH FARM MANAGEMENT

Apurba Bhuyan[1] and A.K.Panigrahi[2]

Department of Fishery Extension, Faculty of Fishery Science
West Bengal University of Animal and Fishery Sciences, Mohanpur Campus, 741252,India

ABSTRACT

The ITK in general is informal in nature and less effort have been made in this regard to document and their subsequent scientific analysis to find out the rationality behind them. An attempt was made to document application of ITK among the fish farmers in the district of North 24 Parganas of West Bengal. The study reveals the use of ITK in fish farm management, which bears scope for further scientific analysis.

The term Indigenous Knowledge (IK) refers to the unique, traditional, local knowledge existing within and developed around the specific conditions of women and men indigenous to a particular geographic area (Grenier, 1998). Das *et al.* (2002) stated Indigenous knowledge as such refers not only to knowledge of indigenous peoples, but to that of any other defined community, while Indigenous Technical Knowledge (ITK) is specifically concerned with actual application of the thinking of the local people in various operations of agriculture and allied areas.

Indigenous Knowledge of farmers has two powerful advantages over outside knowledge are - it has little cost and is readily available. Indigenous knowledge systems and technologies are found to be socially desirable, economically affordable, sustainable and involve minimum risk to rural farmers and producers, and above all, they are widely believed to conserve resources (Grenier, 1998).

The fishers and farmers, through their direct interaction with the fishing and farming situations, have developed lots of working knowledge base and practices. These self gathered knowledge, through experiences and repeated practices, have their unexplored rationality. Albeit the emergence of growing concern of indigenous knowledge worldwide, relevant studies regarding fisheries in this line is less evident. Against this background, the present study has been aimed to document and conduct systematic studies on the indigenous knowledge in fisheries and aquaculture with the following objectives:

- To recognize and document ITK regarding fish farming.
- To short out gender concern in different practices.
- To short out practices according to farmer type.

METHODS

The district 24 Paraganas (N) is selected for investing existing Indigenous Technical Knowledge system as the district is found to be rich in terms of fishery resources along with a good number of fisher folk communities belonging to Scheduled Caste and Scheduled Tribe. Besides, it has good physical facilities like communication, transportation etc. and the fishermen communities in the district are considered to be comparatively more developed by different experts.

Purposive as well as Random sampling techniques with applying lottery method, were adopted for the present study. For selection of state, district and blocks purposive sampling technique was carried out. Two Gram Panchayats were selected randomly out of eight. Out of nineteen villages, only six villages were randomly selected. From each village, a twenty number of respondents selected randomly with a total sample respondents of 120.

Methods followed for data collection in any research work play a very important role, as the methods are main structure of data collection and information exchange (Basu *et al*, 1996). Basu and Biju (1996) have also identified 59 methods going through available literature for the identification of various indigenous knowledge in different perspectives. In the present investigation, interview schedule and Direct observation methods were selected from the set of methods in accordance with the objectives of the study after considering the nature of the fishery resources, type of people and of course the nature of the study.

An interview schedule was prepared with the help of various type of ITK related books, journals etc. Direct observation is systematically observing objects, events, processes, relationship or people and recording these observations. Direct observation is a good way to crosscheck respondents' answers. Information were recorded in notebook. Schematic diagrams, photographs of the materials and indigenous practices were taken.

Categorization of Farmer

The respondents were categorized in to four groups viz. Marginal, Small, Medium and Rich farmers on the basis of their land holding size. This was done as per the convenience of the study and the schedule was developed accordingly. The categorization of fanners are made according to their land holding size which includes total fund possessed by farmer for their livelihood i.e. Agricultural land, Land for fishery activity, home stead land and others.

Table 23.1

Land holding size (Acre)	Farmer type
<1 Acre	Marginal
1-2.5 Acre	Small
2.5-5 Acre	Medium
>5 Acre	Rich

RESULTS AND DISCUSSION

Table 23.2

Sl. No.	ITK found	Purpose of use	Farmer type	Gender concern
1	Provision of shade in culture pond.	To provide shelter to cultured fishes especially during hot summer.	Medium farmers.	Male.
2	Application of raw cow-dung to fishpond.	To enhance natural feed production.	Small and medium farmers.	Male and female.
3	Application of Cattle urine in fishpond.	To control algal bloom.	Small and medium farmers	Male and female.
4	Placing fish feeds in gunny bags immersed in pond water.	To feed fishes and to avoid wastage of feed.	Small and medium farmers.	Male and female.
5	Application of Banana stem pieces in to fishpond.	To purify pond water in occurrence of foul smell, gas bubbles from pond water and to combat illegal fish poisoning.	Small Medium and rich fish farmers.	Male
6	Broadcasting of ash in to fishpond.	As prophylactic measure against fish diseases.	Small and Medium farmers.	Male and Female.

INTERPRETATION OF FINDINGS

Provision of shade in culture pond

Long and broad leaves of Coconut tree *(Cocos nucifera)* or Khajur tree *(Phoenix sylvestris)* are tied with 2-3 bamboo poles. The poles are fixed to bottom of the pond. The leaves are tied in such a position that float or partially submersed into the pond water. The structure is made to provide shelter to cultured fishes specially to overcome hot summer. Locally available materials are used and easy for maintenance.

Application of raw cow-dung to fishpond

Raw cow dung is broadcasted in the fishpond to produce natural fish feed and tempering of water for proper growth of fish. But according to scientific interpretation, Cow dung serves as an ideal source of organic manure, which enhances plank tonic feed items for fishes in pond.

Application of Cattle urine in fishpond

Farmers spray Cattle urine in occurrence of algal bloom in the culture fishpond. This is a simple method and cattle urine may control algal bloom.

Placing fish feeds in gunny bags immersed in pond water

Feedstuff such as Mustard oil cake, Mohua oil cakes are put in to gunny bag and the mouth of the bag is tied up with a jute rope tightly. One to three such bags with feeds are now kept submersed under water near pond bank in 2-3 corners of the pond. Sometimes the bags are tied up with Bamboo pole and fixed to bottom such that the bags remain under water. The feeds are released from the pores of the bags slowly. Thus it checks wastages of feeds.

Other feedstuff includes wheat flour mixed with water, boiled rice, leaves of Pumpkin and Banana. The later two items are specially applied as feed for Grass carp. These items are applied by broadcasting to pond water. Method is simple and may be efficient to feed the fishes.

Application of Banana stem pieces in to fishpond

Banana stems are cut into small pieces on the bank of the fishpond and sprayed into the fishpond. Farmers follow this practice widely when the there is occurrence of fish mortality without any external signs, occurrence of foul smell and gas bubbles from pond water. It is also practiced to purify pond water when farmer's pond has been poisoned for poaching by anonymous. Method is easy, material is readily available, cheap and Banana pseudo-stem may act upon harmful gases and toxicants in pond water.

Broadcasting of ash in to fishpond

Paddy straws are burnt and the ash of it is broadcasted into fishpond or sometimes heaped in one corner of the cultured pond. The practice is used as prophylactic measure to check fish disease in cultured pond. The practice involves least cost and materials are readily available. Applying burnt ash of paddy straw may keep pond water quality congenial for fish growth.

The paper has indicated the rich extent of Indigenous Technical Knowledge and practices in fisheries in the study area. Farmers and fishers have long been bearing up these knowledge and practices and most often use them in their farming situations. The study also reveals involvement of women in different fishing and farming activities. Their participation call for women specific approaches in fisheries technology development, which would stand by, the growing concern of women empowerment.

Formation of Indigenous Knowledge Confers nationwide and its effective network to the grass root level is the need of hour. We have the opportunity to conserve and validate Indigenous Technical knowledge from the indigenous people to make them more productive. It needs back up from the scientific analysis and interpretations. However, the outcome of these interpretations needs to be generalized and disseminated to its end users. Mere documentation and validation of ITK will not be a wise step if the same aren't returned to its owners.

REFERENCES

1. Basu, D. 1996. Participatory Monitoring and Evaluation of Watershed Programme Paper presented in the training course of Watershed development team, SIPRD, Kalyani.

2. Basu, D. and Biju, K. 1996. Indigenous Knowledge system: Some basic dimension. An occasional paper submitted to the Department of Agricultural extension, B.C.K.V., Mohanpur.

3. Chander, M. and Mukherjee, R. 1994. Traditional Agricultural and Animal Husbandry practices for Sustainable Agriculture in Kumoun hills of UP., Journal of Rural Development, 13(3): 1994. pp 443-449.

4. Das, P., Das, S. K., Arya H.P.S., Reddy G.S. and Mishra, A. 2002. Inventory of Indigenous Technical Knowledge in Agriculture. Mission Mode Project on Collection, Documentation and Validation of Indigenous Technical Knowledge Document-I, ICAR.pp.3-8

5. Grenier, L. 1998. Working with Indigenous Knowledge-A guide for Researchers Ottawa: IDRC. In. Paul, S. and Ramanathan, A. 2002. Conservation of Biodiversity Through Indigenous Knowledge. Kurukshetra. Vol.5 No.3 Jan-2002, pp.3-7

6. Nogard, R. D. 1984. Traditional Agricultural Knowledge: Pest performance, Future prospects and institutional implications, American Journal of Agricultural Economy, 66: pp 874-878.

7. Pattnaik, B. C. 2000. Marketing of Traditional Value-added fish products A profitable venture. Fishing chimes, September 2000, Vol. 20 No. 6; pp 52-54.

8. Ray, G. L. 2004. Research Methods in Social Sciences and Extension Education. Kalyani Publishers. Pp.236-237

CHAPTER 24

On The Function of The Labyrinthine Organ of *Macropodus Cupanus*

G.M. Natarajan, N. Parthi*, P. Esther Joice, D. Mallikaraj, P. Palanisamy N. Bhuvaneshwari, G. Sasikala, and S. Binu Kumari**

Post Graduate and Research Department of Zoology, Government Arts College (Autonomous) Coimbatore - 641 018.

**Dublin-9, Ireland*

***Chikanna Govt Arts College, Tirupur- 641 602.*

ABSTRACT

Amputation of the labyrinthine organ of *M.cupanus* increased the utilization of oxygen from water and air. The role of the labyrinthine organ in enduring aerial conditions is discussed.

Key words : *Labyrinthine organ, M.cypanus, Amputation*

INTRODUCTION

The respiratory function of the air-breathing fishes has been extensively investigated in air-breathing fishes (Raveendran, 2000). However, the extent of participation of labyrinthine organs in anabantids in bimodal gas exchange remain speculative. However, Natarajan (1972) has shown by amputation experiments that the labyrinthine organ of *Anabas scandem* participates primarily during bimodal gas exchange. The aim of the present investigation is to assess the role played by the labyrinthine organ of *Macropodus cupanus* in gas exchange mechanism of the fish.

MATERIALS AND METHOD

Samples of *Macropodus cupanus* (8.62-21.67g) were held in bimodal respiratory chambers and the O_2 uptake of the fish in air, water, or in air and water were studied separately

Corresponding Author

Prof. Dr. G.M. Natarajan Department of Zoology Government Arts College (Autonomous), Coimbatore - 641 018.

adopting the procedures and calculations described by Natarajan (1972). The labyrinthine organs were amputated in one batch of fishes. The other batch served as control. For amputating the labyrinthine organs, the fishes were anesthetized with Ms 222 and the labyrinthine organs was removed by cutting off at the base of its origin.

RESULTS AND DISCUSSION

While the control fishes consume 125.38 ± 64.77 mlO_2 / kg / hr of O_2 from air, the amputated fishes show a value of 127.13± 24.40 mlO_2/kg/hr (Table 24.1). However, there was slightly increased utilization of O_2 from water, when compared to the control and amputated fishes under aquatic conditions. The O_2 uptake from the water of the control and amputated fishes were not statistically significant under bimodal conditions. However, there is a significant increase in the O_2 uptake of amputated fishes from air under bimodal conditions when compared with the control fishes. It is apparent that in the amputated fishes, increased dependence on aerial respiration during bimodal breathing is very evident. Further, amputated fishes surface more frequently (38/hr) composed to the controls (19/hr). The survival times of the fishes without labyrinthine organs, out of water ranged from 4-6 hr compared to 8-12 hr in the case of normal fishes, indicating a role of the labyrinthine organ in enduring the aerial conditions.

Table 24.1

Effect of amputation of the labyrinthine organ on the O_2 uptake of M. *Cupanus* (8-22g) mlO_2 /Kg/hr± SD) N=12

O_2 up take in air		O_2 up take in water		Bimodal O_2 uptake				Total	
				O_2 up take in air		O_2 up take in water			
Control	Amputated	Control	Amputated	Control	Amputated	Control	Amputated	Control	Amputated
125.38	127.13	41.34	57.33	39.98	42.58	93.75	149.16	133.66	191.74
±64.75	±24.40	±6.78	±12.65	±16.82	±9.21	±33.62	±27.49	±49.54	+25.89
	+1.40		+38.68		+6.50		+59.10		+43.45
	t=1.75		t=5.72		t=0.69		t=6.28		t=5.33
	P=N.S		P< 0.001		P=N.S		P < .001		P<0.001

Amputation of the labyrinthine organ amply indicates the selective participation of this organ in the gas exchange of M. *cupanus*. Under completely aerial conditions, the amputated and control fishes showed no significant difference in gas exchange. It is therefore obvious that the labyrinthine organ exercises a negligible role in gas exchange on exposure to air. Under bimodal conditions, the dramatic increase in the aerial respiration gives a clue that the labyrinthine organs are more useful for serving the respiratory needs only at this phase.

Normally paddy fields and other deoxygenated waters are the habitats of these fishes. Effective participation of the labyrinthine organs will definitely enhance the survival value of this species under adverse ecological niches. The limited participation of labyrinthine organ in the aerial conditions is perhaps a consequence of desiccation of its respiratory surface. The greater frequency of surfacing of amputated fishes indicate that the loss of air-breathing organs might have limited the retention of air in the labyrinthine organs. Under these circumstances prolonged stay of the fish under water is not possible. As a result, dependence on surface gulping is resorted.

REFERENCES

Natarajan, G.M. 1972. On the respiration of *Anabas scandens*. Thesis submitted to I.C.A.R. New Delhi.

Raveendran. S. 2000. Studies on the bimodal respiration of *Mystus gulio*. Ph.D. thesis submitted to Bharathidasan University, Trichy.

Normally paddy fields and other deoxygenated waters are the habitats of these fishes. Effective participation of the labyrinthine organs will definitely enhance the survival value of this species under adverse ecological niches. The limited participation of labyrinthine organ in the aerial conditions is perhaps a consequence of desiccation of its respiratory surface. The greater frequency of surfacing of amputated fishes indicate that the loss of air-breathing organs might have limited the retention of air in the labyrinthine organs. Under these circumstances prolonged stay of the fish under water is not possible. As a result, dependence on surface air is required.

REFERENCES

Natarajan, G.M. 1972. On the respiration of *Anabas scandens*. Thesis submitted to I.C.A.R. New Delhi.

Raveendran, S. 2000. Studies on the bimodal respiration of *Mystus gulio*. Ph.D. thesis submitted to Bharathidasan University, Trichy.

CHAPTER 25

A Study on the Fecundity of Some Freshwater Fishes of India, with a Note on a Concept of Comparative Fecundity

Pradip Kumar, V.C.Gu...

Department of Zoology, B.S.A. College

Department of Zoology, R.B.S. College

ABSTRACT

Quantitative fecundity data of *Wallago* ... has been collected and analyzed. The comparison of ... between the numbers of ova to the total weight of fish has been elucidated in co...

Keywords: *Fecundity, Freshwater fishes*

INTRODUCTION

The present investigation deals with th... problem because of its direct bearing on fish p... successful management and exploitation of in... on the limits of fishing of various species, only... fish is established. Fecundity in fish depends... has to face in its different stages of life hist... produce larger number of ova have to face a h... and those which produce lesser number of eg... parental care, or do not have many enemies i...

Two types of fecundities are recognized... fecundity; the former relates to the number o... the ovary and the latter signifies the number... stages (as described by Vladykov, 1956). Th...

CHAPTER 25

A STUDY ON THE FECUNDITY OF SOME FRESHWATER FISHES OF INDIA, WITH A NOTE ON A CONCEPT OF COMPARATIVE FECUNDITY

Pradip Kumar, Y.C.Gupta and Vishal Tiwari*

Department of Zoology, B.S.A. College, Mathura-281001, Uttar Pradesh, India.
Department of Zoology, R.B.S. College, Agra-282002, Uttar Pradesh, India.

ABSTRACT

Quantitative fecundity data of *Wallago cittu, Ophiocephalus punctatus, Rita rita* has been collected and analyzed The comparison of the ratio between the numbers of ova to the total weight of fish has been elucidated in contrast to its total fecundity

Keywords: *Fecundity, Freshwater fishes, Wallago attu, Channa punctatus, Rita rita.*

INTRODUCTION

The present investigation deals with the fecundity of fishes, which is an important problem because of its direct bearing on fish production and fisheries. It is very important for successful management and exploitation of inland fisheries. Restrictions have to be imposed on the limits of fishing of various species, only after the fecundity and rate of survival of the fish is established. Fecundity in fish depends upon various environmental factors, which it has to face in its different stages of life history. In nature's balance such of the fishes as produce larger number of ova have to face a high mortality at different stages of life history; and those which produce lesser number of eggs are either viviparous or have some sort of parental care, or do not have many enemies in their surroundings.

Two types of fecundities are recognized today: (i) Relative fecundity; and (ii) Total fecundity; the former relates to the number of ova at second and third stages of ripening of the ovary and the latter signifies the number of ripe ova present in the ovary at IV and V stages (as described by Vladykov, 1956). The difference between the relative and the total fecundity is indicative of the loss of eggs through atresia during first to fourth stages of ripening of the ovary.

Some important contributions on the said field of studies are from Alikunhi (1956); Jhingran (1961); Varghese (1973); Rama (1977); Devadoss (1989); Neja (1992); Antony (1997); Coward and Bromage (1999), Mcgurk (2000), and that of Smdhe *et al.* (2002).

MATERIALS AND METHODS

Live fishes were netted and the ripe females were preserved in 10% formalin for studies on fecundities. Before proceeding with the egg count, observations were made on the females as regards their total length in millimeters, and the weight of the fish in grams using a precision balance. Following this the weight of each ovary was observed and its volume was noted by water displacement method. The diameter of the average size of ova was recorded using stage micrometer and ocular micrometer; and ultimately the stage of the ovary was established by examining the running eggs, as also by serial longitudinal sections of the ovary stained by Mallory's triple stain. The ovaries of all the species were marked for identification and the eggs allowed to harden in 10% formalin.

After24 hours the eggs were taken out of the preservative and kept on cotton wool for 2 hours and again wrapped in blotting paper for another two hours. Fresh weight of each ovary was taken on a precision balance and noted separately. One gram of ova were weighted for every fish and kept apart, and actual count of these ova was made. The number of ova per gram weight of ovary was used in computing the total number of ova in each case. The average of these specimens of the species for length and weight groups were taken as standard and recorded in the table of observations.

RESULTS AND DISCUSSION

The productivity or viability of one particular freshwater fish depends upon its comparative fecundity rather than its total fecundity. As will be seen from the observations table, the fish with the highest total fecundity (viz, Wallago attu, 167,000) has actually the lowest comparative fecundity, which is the real indicator of its productivity in competition with other species (taking only fecundity as a factor here). Total fecundity increases with increasing weight and age but comparative fecundity may not.

The fecundities of the three species of freshwater fishes investigated by the present authors vary from 167, 940 in a specimen of *Wallago attu* (weighing 4994 grams) to 12, 000 in *Rita rita* (weighing about 3178 grams). So we may conclude that the number of ova produced is directly proportional to the size of the fish both in length and weight. In order to make a comparative study of fecundity of these fishes, the ratio of the total number of ova present to the weight offish in grams, has been computed for all the species. After taking the ratio of the total number of ova to the weight of fish in grams (i.e., comparative fecundity), we find, that the most prolific *Wallago attu* cited falls to the lowest prolificity (33.6), *Ophiocephalus punctatus* to the middle place (36.05); while *Rita rita* occupies the top place (37.7).

It is of paramount importance that fecundity studies in fishes be quantitative; as qualitative data of the past century are of little use in elucidating the comparative fecundity of fish species as established in the present paper. The tabulated data bring out not only the length and weight of the fish, volume of ovaries, weight of ovaries, stage (international) of ovary, diameter of ova, total number of ova, and colour of ova; but also bring out the important correlations (ratios) between length of fish and volume of ova; the weight of fish and weight of ovaries; the total number of ova and size of fish, (total fecundity); and the all important ratio between the number of ova in total weight of fish (comparative fecundity). The fecundity ratios bring out some important conclusions for fishes specially.

ACKNOWLEDGEMENTS

The authors are grateful to the Indian National Science & Documentation Center (INSDOC), New Delhi for providing the necessary literature.

REFERENCES

Alikunhi, K. H. 1956. Observations on fecundity, larval development and early growth *of Labeo bata* (Ham.), md. J. Fish. 3: 216-29.

Antony, R.B.T. 1997. Fecundity fluctuations in the oil sardine, *Sardinella longiceps.* Ind. J. Fish. 18: 84-98.

Coward, K and Bromage, N. R. 1999. Spawning frequency, fecundity, egg size and ovarian histology in groups of *Tilapia zillii* maintained upon distinct food ratio sizes from first feeding to sexual maturity. Aquat. Liv. Reso. 12.11-22.

Devadoss, P. 1989. Observation on the length-weight relationship, food and feeding habits of spadenose shark, *Scolidon laticaudus.* Ind. J. Fish. 36:169-74.

Jhmgran, V. G. 1961. Studies on the maturity and fecundity of the Gangetic anchovy, *Setipinna phasa* (Ham.). Ind. J. Fish. 8:291-311.

Mcgurk, M. D. 2000. Comparison of fecundity-length latitude relationships between non-anadromous and anadromous sockeye salmon, *Oncorhynchus nerka.* Can. J. Zool. 78: 1791-1805.

Neja, Z. 1992. Maturation and fecundity of mackerel, *Scomber scombrus* (Linn.) in Northwest Atlantic. Acta. Ichthyol. Pisc. 22: 125-40.

Raina, H. S. 1977. Observations on the fecundity and spawning behaviour of *Schizothorax esocinus* (Heckel) from Dal lake Kashmir. Ind. J. Fish. 24: 201-03.

Sindhe, V. R.; Rafiq, M. D.; Satish, S. V. and Kulkarni, R. S. 2002. Fecundity of the freshwater fish, *Channa punctatus* from three aquatic bodies. J. Exp. Biol. *23: 429-32.*

Varghese, T. J. 1973. The fecundity of the rohu, *Labeo rohita* (Ham.). Proc. Ind. Acad. Sci. 77: 214-24.

Vladykov, V.D. 1956. Fecundity of wild speckled trout, *Salvelinns fontinalis* in Quebec lakes. J. Fish. Res. Bd. Canada. 13: 799-841.

CHAPTER 26

Haematological Changes In A Hill Stream Fish Garra *Gotyla Gotyla* (Gray) Infected with Trypanosomes

J.V.S. Rauthan, Geet Rauthan and S.P. Grover
Fisheries Research Laboratory, Department of Zoology D.A.V. (P.G.) College, Dehradun — 248001

ABSTRACT

Infected hillstream fish *Garra gotyla gotyla* revealed the presence of Trypanosomes in the blood which causes conspicuous fall in total erythrocyte count (TEC), packed cell volume (PCV), Haemoglobin content (Hb), MCHC content and rise in total leucocyte cound (TLC), Erythrocyte sedimentation rate (ESR), Mean corpuscular volume (MCV). The differential leucocyte count (DLC) revealed a rise in the small lymphocyte, basophils and monocytes, while the neutrophils, thrombocytes, large lymphocytes and eosinophils fall in the infected fish.

Key Words : *Fish Garra gotyla gotyla, Trypanosomes, TEC, TLC, Hb, PCV, ESR, MCHC,MCV and DLC.*

INTRODUCTION

Presence of tiypanosomes in fishes has been reported from all over the world. Smirnova (1971), Tandon and Joshi (1973, 1977), Joshi (1979), Joshi and Dabrai (1981), Shnan, *et. al.* (1984), Gupta and Agarwal (1985), Gupta and Gupta (1987), Joshi (1989), Sharma and Joshi (1991), and Rauthan (1993). The paper present results on some haematological values obtained from the hill stream fish *Garra gotyla gotyla* infected with trypanosomes under natural condition of Doon Valley.

MATERIALS AND METHODS

Live fishes were collected from the river. Song near Railwala and Lacchiwala. The fishes were brought to laboratory in plastic containers and transferred to glass aquaria for 10 days acclimatization. During acclimatization the fishes were fed on usual diet. For all kind of haematological investigation blood was collected from branchial vein using a microsyring with 24 gauge needle. The fishes were sacrificed and studies on the peseted haematological parameters. Out of these 30 specimens, 12 were found trypanosomes. The studies were made on the selected parameters viz. total erythrocyte count (TEC), Total leucocyte count (TLC), Haemoglobin contents (Hb), Packed Cell Volume (PCV), Erythrocyte Sedimentation Rate (ESR), Mean Corpuscular Haemoglobin Concentration (MCHC), Mean Corpuscular Volume (MCV) and Differential Leucocyte Count (DLC) following the standard methods as used Rauthan (1993).

RESULTS AND DISCUSSION

TEC was found lowered in diseased fish as compared to the healthy ones. The fall in diseased fish was 28.0% against the normal healthy fish (2.89 ±0.22x10^6/cmm). The Haemoglobin content (Hb) were also found depleted by 27.8% in diseased fish against the healthy fish (10.4+1.03 gm%).

Table 26.1

Haematological changes in hillstream fish *Garni golyla gotyla* (Gray), All values are mean ± SE for six specimens each.

Parameters	Healthy Fish	Infected Fish
TEC (x 10^6/cmm)	2.89±0.22	2.06±0.20
TLC (x 10^3/cmm)	8840.81±120.2	10742.00±205.0
Hb (gm %)	10.4±1.03	7.5±1.06
PCV (%)	32.0±2.0	27.0±1.80
ESR (mm/hour)	1.12±0.49	2.30±0.20
MCV (mm^3)	170.0±10.0	202.0±13.0
MCHC (%)	34.2±1.4	27.3±1.6

Table 26.2

Differential Blood Cell Counts (%)

Large lymphocyte	7.2±0.52	5.3±1.2
Small lymphocyte	25.51±2.3	30.0±3.0
Monocyte	1.31±0.56	2.40±0.30
Neutrophil	20.2±3.5	16.2±2.6
Eosinophil	0.38±0.2	2.2±0.4
Basophil	1.60±0.4	1.90±0.6
Tlirombocyte	42.2	30.3±4.6

The TLC value was found to have increased under disease by about 16.1% against the healthy fish value of 8840.81±120.2 x 10^3/cmm. The PCV value also shewed a fall of about 15.6%. The MCHC was found lower in infected fish as compared to the healthy fish 20.1%. The MCV value had also rise in the infected fish by about 15.8% and the ESR value rose by about 51.3% as compared to the healthy fish values. The differential cell counts also revealed conspicuous change. The number of small lymphocyte, bisophils and eosinophils rose due to infection, but the number of thrombocytes and neutrophils fall in the diseased fishes, as compared to the healthy fishes. The number of large lymphocytes, like those of the neutrophils also fell by a little margin in infected fish as compared to the health ones (Table 26.2). The monocytes were found higher in infected fishes as compared to the healthy fishes.

Parasitic infections have been found to destroy eiythrocytes and cause anemia (Ven Brand 1973). Decrease Hb percentage. high value of ESR and fall in PCV value is infected fishes in the present study are clear indications of anaemia. The present study tends support to Englel and Davis' (1961), Smirnova (1971), Tandon and Joshi (1973), Joshi (1979), Joshi and Dabral (1981) and Joshi (1989).

As many as ten species of fresh water teleosts having haemoflagellate trypanosomes in their blood fully revealed that trypanosomiasis cause definite alterations in various blood component TEC, Hb and PCV values fall due to these haemaflagellates while the TLC usually rise (Tandon and Joshi 1973) but Joshi and Dabral (1981) reported the TLC value fell sharply and significantly between 60-100% under heavy infection. The RBC, WBC, Hb and PCV values were found to have depleted in cat fish. Kumar, et. al. (1984) reported TLC and ESR values are highest in *Schizothorax rachardsonii* severally suffering from black spot disease. Smimova (1970) in the perch *Lota lota* infected with trypanosomes noted decrease TEC, and Hb but increased number of leucocytes.

Similarly a recent report by Gupta and Gupta (1990) following experimental inoculation of trypanosomes in the fresh water shark as *Wallago allu* also found severe anaema 39 days post infection, where TEC had fallen from 2.4 to 1.6 x 10^6/cmm and Hb from 11.4 to 8.0 gm%. The differential blood cell counts also revealed noticeable changes. The small lymphocytes, basophils, monocytes and Eosinophils rose in the infected fish (Table 26.1), while the number of neutrophils and thromocyte and large lymphocyte fall (Table 26.2).

ACKNOWLEDGEMENTS

The authors wish to express theri indebtedness to Dr. J.K. Srivastava, Head, Department of Zoology, D.A.V.(P.G.) College, Dehra Dun for valuable suggestions and critical reading of the manuscript.

REFERENCES

Engel, DW and Davis, EM, 1964 : Relationship between activity and blood composition in certain teleosts. *Copeia,* 31 : 586-587.

Gupta, N. and Gupta, D.K., 1986 : Trypanosome infectivity and changes in the glucose level of two fresh water fishes, *J. Parasitology* 10(2):213-213.

Joshi, B.D., 1979 : Hyperlactemia in the fish blood infected wtih trypanosomes. *Sci. Cult.*, 45(8) : 320-321.

Joshi, B.D. and Dabral, R. 1981 : Some haematologica chagnes in fresh water cat fish *Heteropneustes fossilis* infected with trypanosome. *Proc. Ind. Acad. Sci.*, 90 : 295-301.

Joshi B.D. 1989 : Physico-pathological studies on the blood of few hill stream Teleosts. *Proc. Natt. Symp.*: 127-137.

Kumar, A., and Singh, H.R., 1984 : Haematological changes in cold water fish *Schizothorax plogiostomus* (Heckel), infected with black spot disease. *Ind. J. Parasit*, 8(1): 129-131.

Rauthan, J.V.S., 1993 : *Studies of the haematological variations in some fishes of Doon Valley.* Thesis submitted to H.N.B. Garhwal University, Srinagar (Garhwal), 1-221.

Radhakrishnan, S., Nair, N.B., and Balasubramanian N.K., 1984 : *Gymnorhycus gigas* plerocercoid infection of the lvier of *Diodon hystrix* (Pisces : Diodontidae) II. Haematological changes infected fish. *Fish and unwelt*, 13 : 27-39.

Smrihova, L.I.; 1970 : Trypanosome in the blood of *Lota lota* L. (Tryponosoma lotai sps. Nov.) *Parasitologiya* 4 : 296-297.

Tandon, R.S. and Joshi, B.D., 1973 : Studies on the physiopathology of blood of fresh water fishes infected with two new forms of trypanosomes. *Z. wies Zool. Leipzg.*, 185 (314) : 207-221.

Von. Brand, T. 1973 : *Biochemistry of Parasites.* 2nd Ed. Academic Press, New York.

CHAPTER 27

Emersion Stress Changes the Free Amino Acid Content of Blood of *Macropodus Cupanus*

G.M. Natarajan, N. Parthi*, D. Mallikaraj, P. Palanisamy, P. Esther Joice, N. Bhuvaneswari, S. Binu Kumari and G. Sasikala**

Post Graduate and Research Department of Zoology, Government Arts College (Autonomous) Coimbatore - 641 018.
**Dublin-9, Ireland*
***Chikanna Govt Arts College, Tirupur- 641 602.*

ABSTRACT

Emersion increased the FAA content of blood of *M. cupanus*. The role of FAA in enduring emersion stress is discussed.

Key words : *FAA, Emersion, M. cupanus*

INTRODUCTION

Air-breathing fishes are known to undertake overland forays. Naturally, fishes adopting these terrestrial excursions must have developed some biochemical adaptations for water deprivation. The object of the present investigation is to study the emersion stress induced changes in the FAA composition in the blood of *M. cupanus*.

MATERIALS AND METHOD

Groups (20-25g) of *M. cupanus* were air exposed by placing individually in 0.51 capacity covered plastic boxes containing 100ml of filtered tap water. The relative humidity (RH) inside the box was 85-90% RH and the temperature 28° ± 1°C. The plastic boxes containing the fishes were exposed to air for 8 hours. The blood for FAA estimation was collected and deprotinized in 80% alcohol and centrifuged. The FAA were quantified by two dimensional

Corresponding Author

Prof. Dr. G.M. Natarajan Department of Zoology Government Arts College (Autonomous), Coimbatore-641 018.

paper chromatographic technique described by Block *et al*., (1958). The optical density of each sample was read in a Spectronic 20 (Bauch and Lomb) Colorimeter at 550 filter.

RESULTS AND DISCUSSION

Normally 16 amino acids are detectable in the blood of *M. cupanus* (Table 27.1). 4 hr emersion register a rise of 52% in the total FAA content. 8 hr. air-exposure show highest increase of nearly 69% in total aminoacid concentrations over normal value. Excepting ornithine and. lysine, all other aminoacids register a continuous increase and reaches the maximum values at 8 hr exposure.

Table 27.1

Emersion induced changes in the FAA levels in thc blood of *Macro podus cupanus* (Values expressed as Mg/ml x + SE from 3 observations)

Amino Acids	Normal	4hr air exposure	% increase	8hr air exposure	% increase
Alanine	12.30±1.22	15.62±3.16	+26.99	18.20±6.21	+47.97
Arginine	19.78±4.21	26.71±8.20	+35.04	30.31±16.20	+53.24
Asparagine	27.60±10.30	36.82±12.81	+33.41	40.80±13.18	+47.83
Aspartic acid	25.24±4.76	37.20±10.66	+47.39	42.60±15.20	+68.78
Glutamic acid	30.26±5.38	42.68±6.40	+41.04	56.40±8.00	+86.39
Glycine	19.74±2.69	24.88±3.50	+26.04	30.16±10.12	+52.79
Histidine	21.35±4.60	31.26±10.38	+46.42	33.60±9.80	+57.38
Leucine	6.20±0.78	9.28±1.16	+49.68	11.80±4.21	+90.32
Lysine	17.35±2.88	40.60±14.96	+134.55	39.06±11.32	+67.88
Methionine	5.24±0.80	7.60±0.89	+45.04	10.21±0.70	+94.85
Ornithine	44.54±12.61	62.50±16.62	+40.32	60.18±13.54	+35.12
Phenylalanine	4.60±1.76	9.36±2.70	+103.48	11.63±1.55	+ 152.83
Proline	-	-	-	-	-
Serine	18.45±4.67	30.16±14.60	+63.47	33.18±16.21	+79.84
Threonine	7.15±0.92	10.16±2.44	±42.10	14.50±2.53	+ 102.80
Tyrosine	3.21±0.65	5.82±0.72	+81.31	7.60±1.20	+ 136.76
Valine	11.56±2.70	26.54±9.80	+129.59	29.00±11.60	+ 150.87
Total	274.53±60.93	417.19±119.0	+51.97	463.23±141.6	+68.74

FAA play a very important role in osmoregulation. In the Sea-anemone *Actinia equina* (Navarro and Ortega, 1984) the FAA contribute significantly to overcome the problems of osmoregulation during air-exposure. Air exposure increases both the essential and non-essential amino acids. While the increase of the non-essential amino acids in acceptable, the increase in the essential amino acids calls for through probe. Probably emersion or desiccation stress degraded the tissue proteins. Also, the significant increase in the FAA phenylalanine, valine, tyrosine and threonine at 8hr needs further experimentation. Whether they play any role in the detoxification process or osmoregulatory adjustments or battering capacity of the blood remains to be seen.

REFERENCES

Block. R.J.. Durrum. E.L. and Zewig, G. 1958. *Paper chromatography. A. Manual of Paper Chromatography and paper Eiectrophoresis.* Academic Press. New York.

Navarro. E. and Ortega, MM 1984. Free ammoacid accumulation from glucose during air-exposure and anoxia in the sea-anemone. *Actinia equina. Comp. Biochem. Physiol.*, 78B. 199-202.

CHAPTER 28

Length-Weight Relationship and Fecundity of Cyprinid Fish *Semiplotus manipurensis* from Manipur

Laishram Kosygin and Waikhom Vishwanath*

Loktak Development Authority, Lamphel - 795004, Manipur, India.

**Department of Life Sciences, Manipur University, Canchipur - 795003, India.*

ABSTRACT

A preliminary study on the biology of *Semiplotus manipurensis* was carried out during May, 1997 - March, 1998. As many as 304 specimens of the species ranging from 60.0 to 235.0 mm in total length were taken for the present study. The length-weight equation for the fish was calculated as Log W = 3.1508895 Log L - 5.1984116. The average value of relative condition factor "Kn" was found to be 1 (one) indicating a general good condition. The regression equation between the length and weight was calculated as Y = - 90.786444 + 1.0268897 X. Fecundity varies from 18298 to 20573 which shows that the fish is a prolific breeder.

Key Words: *Length-weight relationship, Fecundity, Semiplotus manipurmnsis, Manipur.*

INTRODUCTION

The natural aquatic ecosystems of Manipur are subjects to considerable stresses due to changes in the environment and various human activities. These lead to decrease in fish population, which forms one of the main components of aquatic ecosystem. Therefore a need has arisen to take up necessary steps so as to conserve the vast and diverse fish genetic resources on one hand and to rational and efficient utilization of fish stock on the other hand. In Manipur, efforts have been made to study the biological aspects of some economically important fishes (Sharma, 1989). Hora (1921) pointed out that fish is the only animal protein in the diet of the people of Manipur.

Semiplotus manipurensis Vish. and Kosygin (2000) is a delicious carp and is so far known only from Ukhrul district of Manipur. Due to difficult hill terrain and lack of proper means of communication from the fishing areas to the fish landing centers, fishermen smoke the fish and bring to Ukhrul market where it fetches high prices. However, there is no report on the biology of the fish. The paper presents a brief information on the length-weight relationship, relative condition factor and fecundity of *S. manipurensis.*

MATERIALS AND METHODS

Specimens of *S. manipurensis* were collected from the Challou river during the period from May, 1997 to March, 1998. As many as 304 specimens of size ranging from 60.0 to 235.0 mm in total length were collected. In all cases the samples were preserved in 5% formalin solution immediately after collection. Care was taken not to damage the caudal fin while collecting the fish. Length of the fish was measured by using digital-dial slide caliper and a top pan electronic balance measured total weight of the fish. Total length was measured from the tip of snout to the longest tip of caudal fin. The mathematical relations between length and weight were calculated following Le Cren (1951).

Ten matured female specimens of *S. manipurensis* ranging from 210.0 to 241.5 mm in total length, were studied. Fecundity was estimated by gravimetric method (Khanna, 1991). The diameter of eggs was measured using a digital-dial slide caliper.

RESULTS

Length-Weight Relationship: The length-weight relationship of *S. manipurensis* was calculated from measurement of 304 specimens ranging from 60.0 to 235.0 mm in total length and from 2.52 to 187.10 gm in total weight. The length-weight equation for the fish is expressed as follows:

$$\text{Log } W = 3.1508895 \text{ Log } L - 5.1984116$$

The correlation co-efficient 'r' between total length and total weight of the body is calculated as 0.955 indicating a very high degree of correlation.

The regression equation between the length and weight of the fish was calculated as $Y = -90.786444 + 1.0268897\,X$

The co-efficient of condition factor (k) of the species was found to vary from 1.06 (in 65.14 mm in TL) to 1.49 (in 182.10 mm in TL) with an average of 1.34. The relative condition factor (Kn) based on the observed weight and corresponding calculated weight data (by eliminating the effects of length and its other correlated factors) fluctuates from 0.85 to 1.13 with an average value of 1. The minimum and maximum values were obtained at size 167.9 mm and 86.36 mm respectively.

The graph assumes a exponential curve when the average weights were plotted against the mean length (Graph 28.l), but when both variables are depicted in terms of logarithms (log weight-length) they assumes a straight line (Graph 28.2).

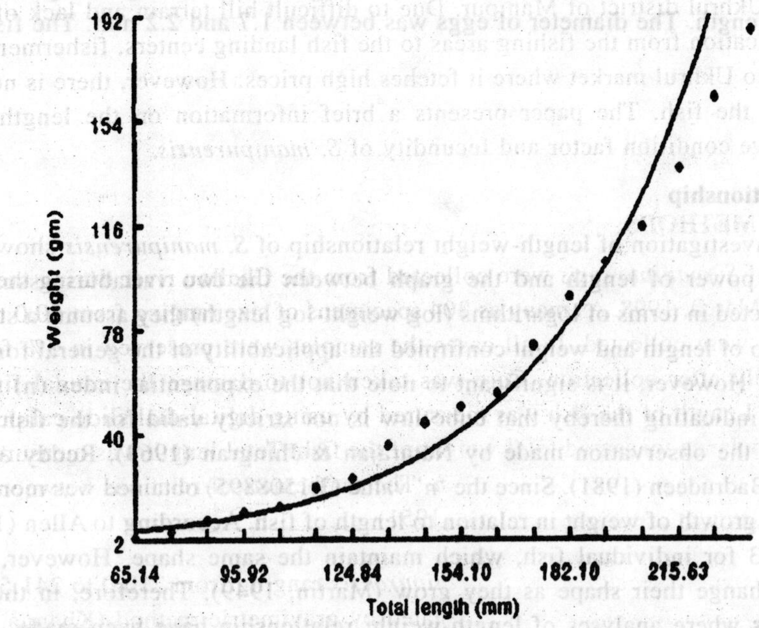

Graph 28.1. Length-weight relationship of *S. manipurensis*.

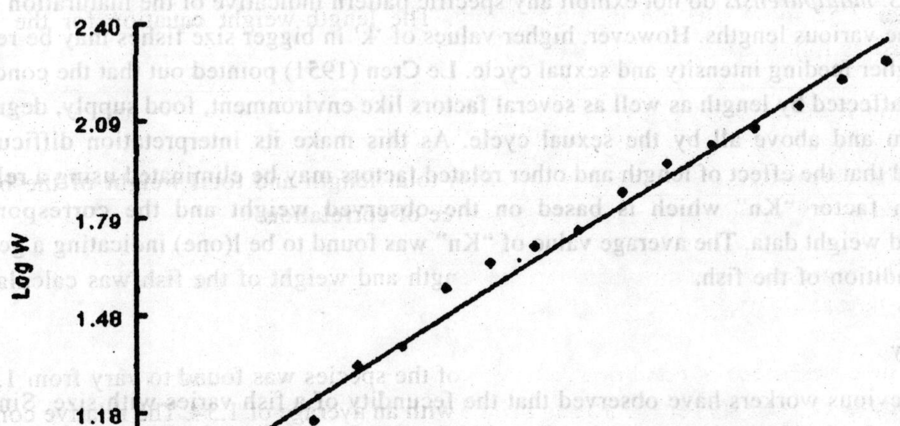

Graph 28.2. Relationship between Log L and Log W of *S. manipurensis*.

Fecundity: Mature females of *S. manipurensis* of stage IV and V ovaries were used for the study. The mature ovary of the fish contains large orange yellowish eggs. The total number of mature ova per female varied between 18298 and 20573 in the length range of 210.0 - 241.5 mm in total length. The diameter of eggs was between 1.7 and 2.2 mm. The fish is a prolific breeder.

DISCUSSION

Length-Weight Relationship

The present investigation of length-weight relationship of *S. manipurensis* shows that weight increases as power of length and the graph between the two variables assumes a curve, but when depicted in terms of logarithms (log weight-log length) they assume a straight line. This relationship of length and weight confirmed the applicability of the general formula, $W = a L^n$ to the fish. However, it is significant to note that the exponential index (n) shows estimation above 3, indicating thereby that cube law is not strictly valid for the fish. This finding is similar to the observation made by Natarajan & Jhingran (1963), Reddy & Rao (1992) and James & Badrudeen (1981). Since the 'n' value (3.1508895) obtained was more than 3 the fish has a good growth of weight in relation to length of fish. According to Allen (1938), 'n' value equals to 3 for individual fish, which maintain the same shape. However, most species of fish do change their shape as they grow (Martin, 1949). Therefore, in the vast majority of instances where analyses of length-weight relationship have been made, it has been observed that empirical observation strictly do not conform to this cube law and equilibrium constant frequently shows certain variability around 3. In *S. manipurensis* body seems to become deeper as the fish grow (depth of body in percent of standard length increase as the fish grow). This might be one reason; the fish has a better growth of weight in relation to length of fish. The value of the co-efficient of condition (k) or the general well being of *S. manipurensis* do not exhibit any specific pattern indicative of the maturation cycle among the various lengths. However, higher values of 'k' in bigger size fishes may be related to the higher feeding intensity and sexual cycle. Le Cren (1951) pointed out that the condition factor is affected by length as well as several factors like environment, food supply, degree of parasitism and above all by the sexual cycle. As this make its interpretation difficult he suggested that the effect of length and other related factors may be eliminated using a relative condition factor "Kn" which is based on the observed weight and the corresponding calculated weight data. The average value of "Kn" was found to be l(one) indicating a general good condition of the fish.

Fecundity

Previous workers have observed that the fecundity of a fish varies with size. Sine the fecundity of only 10 specimens ranging from 210.0 - 241.5 mm in standard length were studied during the present investigation, which is considered insufficient to establish any definite relationship with the length and weight of the fish, no attempt is made on these line. However, in the present investigation fecundity varies from 18298 to 20573 which shows that the fish is a prolific breeder.

From the above discussion it is clear that *Semiplotus manipurensis* is a healthy delicious carp and prolific breeder. The fish also has a high food value and is sold in the

Fig. 28.1. A view of Chalou River at Chingai, Ukhrul District, Manipur.

Fig. 28.2. ***Semiplotus manipurensis*** **(Vish. & Kosygin)**

Ukhrul town from where it is so far reported (Kosygin & Vishwanath, 2003). Considering the importance of the fish, proper attempt should be taken up using scientific techniques to increase the stock of this fish.

ACKNOWLEDGEMENTS

We are grateful to Shri R.K. Jayanta Kumar Singh, Agricultural Officer, Department of Agriculture, Government of Manipur and Lewis Horam of Tusom Village for their valuable help in collecting the specimens of *Semiplotus manipurensis.*

REFERENCES

Allen, K.R. 1938. Some observation on the biology of the trout *(Salmo trutta)* in Windermere. *J. Anim. Ecol.,* 7: 33 - 349.

Hora, S.L. 1921. Fish and Fisheries of Manipur with some observation on those of Naga Hills. *Rec. Indian Mus.* 22(3): 165-214, 4 pls.

James, P.S., and B.R. Badrudeen, 1981. Biology and fishery of silverbelly *Leiognathus dussumieri* (Valenciennes.) from Gulf of Mannar. *Indian J. Fish.* 28: 154 - 182.

Khanna, S.S. 1991. An introduction to fishes. Central Book Depot, Allahabad (3 rd Ed.) 530 pp.

Kosygin, L. and W. Vishwanath, 2003. Prospect of sustainable development of fisheries in the hill streams of Manipur. *J. North Eastern Council,* 23 (3): 18-26.

Le Cren, E.D. 1951. The length-weight relationship and seasonal cycle in gonad weight and condition in the Parch *(Perca fluviatilis). J. Anim. Ecol,* 20: 201 - 219.

Martin, W. R.I 949. The mechanics of environmental control of body form in fishes. Univ. Toranto Stud. Biol. (Publ. Ont. Fish. Res. Lab. 70) 58: 1-91.

Nataranjan, A.V. and Jhingran, A.G. 1963. On the biology *of Catla catla* (Ham.) from the river Jamuna 29(33): 326 - 355.

Reddy, Y.S. and M.B. Rao 1992. Length-weight relationship and relative condition of *Puntius sophore* (Hamilton-Buchanan) from Lake Hussain Sagar, Hyderabad, India. *J. Indland Fish Soc. India.24(1):* 22 - 25.

Sharma, M.G. 1989. Study on the hill stream fishes of Manipur with special reference to eco-biology of *Schizothorax richardsonii* (Gray). Ph.D. Thesis, Manipur University, 420 pp.

Vishwanath, W. and L. Kosygin. 2000. Fishes of the cyprinid genus *Semiplotus* Bleekeı 1859, with description of a new species from Manipur, India. J. Bombay nat. His.Soc. 97(1): 92-102,1 pl.

CHAPTER 29

STUDY ON THE LENGTH-WEIGHT RELATIONSHIP AND POPULATION STRUCTURE OF CHOCOLATE MAHSEER-*NEOLISSOCHEILUS HEXAGONOLEPIS* (MCCLELLAND) IN 'SHALLY LAKE' IN ARUNACHAL PRADESH.

Boni Amin Laskar, Azen Pujen, *Pronob Nath

Senior Research Fellow, NRCCWF (ICAIR) Project, Fisheries Experimental Farm-IDULI, Lower Dibang Valley District - 792110, Arunachal Pradesh.

**Deputy Director , Department of Fisheries, Govt. of Arunachal Pradesh*

ABSTRACT

"Mahseer" the premier sport fish of the country which was once the major fishery of the State of Arunachal Pradesh is on the decline. Among the three varieties available in this region, *Neolissocheilus hexagonolepis* (Chocolate mahseer synonyum of *Accrosocheilus hexagonolepis*) is in the verge of extinction and has been regarded as a threatened species of this region of the Himalayas.

As a step towards conservation measure, the state department has stocked the fishes from time to time in 'Shally Lake' near Roing of Lower Dibang Valley District of Arunachal Pradesh. In order to asses the pattern of growth of this fish in the confined water systematic studies including the studies on population dynamics has been carried out in the monsoon seasons. The studies showed that the fishes has adapted to the confined condition of the lake and the growth pattern is negatively allometric while the average condition factor being 1.005 showed the well-being of the fishes.

The paper deals with the result of the population studies of the different age groups to focus the importance of mahseer and its future prospect to form a component in the culture fishery.

Key Words : *Length-Weight relationship, Lake Chocolate mahseer Allometric growth and natural population.*

INTRODUCTION

Studies on the relationship between length and weight of fishes constitute an important tool in fish Biology. Nevertheless such study are usually undertaken to know whether variations from expected weight for the known length groups are the indicators for fatness, well-being, gonadal development with relation to the environment. (Beckman, 1948; Lecren, 1951; Bagenal. 1978) and fluviatile condition of their habitat.

Length-Weight (L-W) relationship of fishes has been developed through the use of general equation of $W=CL^n$. The general expectation is that the weight increase as the cube of the length (Rousenfell and Everhart.1953 : Lagler, 1956; Brown. 1957). As the fishes passes through several stages the simple cube law docs not hold good through out the life span and equilibrium constant shows certain variations (Martin, 1949) in the growth pattern of fish.

Interestingly, several works had been done on the L-W relationship of commercial freshwater fishes from different water bodies in India (Srivastava and Singh. 1964 : Mitra and Das, 1967 : Jhingran, 1968 : Venkateswarlu and Bannerjee. 1971 Talukdar and Das. 1975 : Pathak, 1976 : Sen, 1982 : Kar and Barbhuiya. 2000 : Patgiri *et.al.*, 2001); but hardly any record is available on Chocolate mahseer in this regard.

PLACE OF EXPERIMENT AND LOCATION

The study was carried out at the shally lake in Lower Dibang Valley District of Arunachal Pradesh lying between 28°15 X and 95° 45' EL located at an altitude of 395m (msl). The lake is of tectonic origin and perhaps created during the year 1950 with water area of 2 (Two) hector. The fishery activities were taken up by the department of fisheries, Govt. of Arunachal Pradesh from the year 1970-71 onwards with provision of sluice gate to check the outflow of water at the outlet. The IMCs were initially stocked and during 1994-95 fires of mahseer (Chocolate mahseer) were thereafter stocked. The lake presently attaches great importance not only for composite fish practices but also for tourist attraction.

MATERIALS AND METHODS

Chocolate mahseer were the principal materials for the study. The fishes has been collected by using dragnet (mesh size 5mm). Altogether, 131 number of Chocolate mahseer (Length group 70-285 mm) were collected from the lake to undertake the studies. Total length (mm) of all the specimens were measured from tip of snout to longest axis of caudal fin, through measuring board. Body weight (g) of each individual fish was also recorded along with the corresponding length.

The L-W relationship was established by fitting in the logarithmic form of general equation W=CL^n: Where L measured length. C= Constant, n=Equilibrium and W= expected weight. The loganthmic form of the equation is Log W= LogC$^+{}_n$ Log L. Log C was calculated by the following formula :

$$Log\,C = \sum \frac{LogW(\sum LogL^2) - \sum LogL * \sum(LogL * LogW)}{N * (\sum LogL^2) - (\sum LogL)^2}$$

The value of 'n' was calculated by the following formula :

$$n = \frac{\sum LogW - N * LogC}{\sum LogL}$$

Condition factor (K) was calculated by the following formula :

$$K = \frac{Observed\,weight}{Expected\,weight}$$

Different statistical methods has been used to evaluate the population structure.

RESULT AND DISCUSSION

1. The present study on L-W relationship indicated a wide variation from the cube law. Such variations has also been reported by several workers (Jhingran, 1952 : Rao and Rao, 1972 : Choudhuri *et. al.,* 1982). The calculated L-W relationship has been LogW =-4.402 ± 2.736 Log L. This shows that the growth although was allometric : yet it could have been isometric (i.e. n=3) and the weight of each fish would have been more from that as observed . The average observe weight was 83.4 g while the average expected weight was 110.27g, Also, the calculated weight per mm of length considering the n value as 3 (three) was 1.738g, but the observed weight per mm of length was 0.42g . Therefore, following the cube law it could be inferred that the body weight of the fishes (N. *hexugonolepis)* would have been increased by 1.32 gm/mm than as observed; if proper nutrition and fertilization been carried out ensuring proper tropic level in the confined ecosystem.

2. The variation in the condition factor may be due to the various factors like environmental conditions, food availability and gonadal maturity as reported by Lacren (1951), Jhingran (1972), Bassirullah (1975) and Dasgupta (1982) . From the present studies it has been recorded that the fish population was with average condition factor =1.005 . The weight of 43% of the population belonging to the length group 130-285(mm) was more than the expected weight. This is perhaps due to the extra- weight of the gonads of the matured/brood fishes. The observation also correlates with the findings of Dasgupta (1982) that even among the members of the population sampled on the same day may show considerable variation in condition factor in relation to length

3. The proportion of different age groups and sexes showed that the population has a definite structure with a broad base when plotted the number of fishes against different size groups (Fig. 29.1); which further reveals the expanding nature of population of chocolate mahseer in the lake.

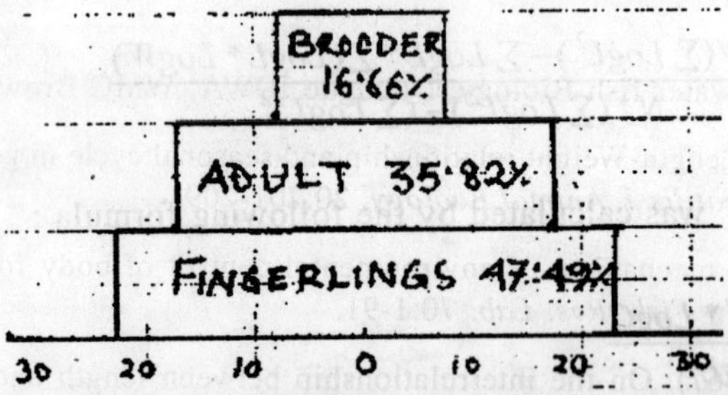

Fig. 29.1 Population structure of Chocolate mahaseer

4. The total sample of fishes has been segregated depending on the maturity status in to Fingerlings (47.49%), Adult (35.82%) and Broaod, fish or Broder (16.66%) and reveals thai the population of chocolate mahseer in the lake has suitably adapted with prominent growth since its introduction in 1994-95.

The above studies lead us to conclude that although the Chocolate mahseer or *N. hexagonolepis* is a semi-rheophilic fish, yet can be suitably culture in the closed environment either as monoculture or as composite culture along with the IMCs

Table 29.1

Details of Length Weight relationship of *Neolissocheilus* hexagonolepis

Source	Number	Length group(mm)	Average observed weight(gm)	Observed weight per mm length	Expected per mm length when n=3	N value	Average condition factor
Shally lake	131	70-285	83.40	0.420	1.738	2.736	1.005

REFERENCES

Beckman, WC(1948) The Length-Weight relationship factors for conservations between standard and total length and co-efficient of condition for seven Michigan fishes. *Trans. Am. Fish. Soc.* 75:237-256.

Bagenal, T(1978) Methods for assessment of fish production in freshwaters. 3rd ed. Blackwell Scientific Publications Ltd. Australia 365p.

Bashirullah, AKN (1975) **Biology of *Lutjanus griseus*(L)** of the Cubagua Island Venezuela Length-Weight. Body length Gut **length** relationship and condition factor. *Bio.Inst.Occanogr. Uni. Orientle* 14(1):101 -107.

Choudhuri, M; Kolekar, V and Chandra, R (1982) Length-Weight relationship and relative condition factor of four Indian major carps of river Brahmaputra *Assam Journal of Inland fishery Society of India* (2): 42-48.

Dasgupta, M (1982) An investigation on some aspects of the Biology of Mahseer from North East India, Ph.D. Thesis North East Hill University, Shillong.

Jhingran, AG(1968) The Length-Weight relationship and K factor of *Gudusia chapra* (Ham.) from the Ganga river system 18:149-163.

Jhingran, VG(1952) General Length-Weight relationship of three major carps of India *Proc. Nisc. India* 18:449-455 .

Lagler, KF(1956) Freshwater fish Biology, Dubugue IOWA, WMC Brown

Lecren, ED(1951). The Length-Weight relationship and seasonal cycle in gonad weight and condition in perch *Perca fluviatilis J. Animal Ecology*, 20:201-209.

Martin, WR (1949) The mechanism of environmental control of body form in fishes, *Univ. Toronto. stu. biol.;58: Ont, Fish. Rvs. Lab.* 70:1-91.

Mitra, B and Das, C(1967). On the interrelationship between length and weight of *Gudusia chapra* (Ham.). *Proc. 54th Indian Sci.Congr. Calcutta* Pt-3, 476 (abstract).

Patgiri. A; Goswami. MM; Kar. D and Barbhuiya, MH(2001) Comparative study on the length-weight relationship and relative condition factors in major and exotic carps in ponds of Guwahati *Indian. J. Environ and Ecoplan.* 5(1):179-180.

Pathak, SC (1976) Length-Weight relationship and condition factor and food study of *Labeo calbasuu (Ham.)* from Loni reservoir (MP). *J. Inland fish. Soc. India* 1:58-64.

Rao, GR and Rao, LH(1972) On the biology of *Labeo calbasu* (Ham-Buch) from the river Godavari. *Journal of inland fishery Society of India* .74-86.

Rousenfell, GA and Everhart,WH (1953) 'Fishery Science' the method and application. John Wiley and Sons, New York.

Sen, N(1982) Studies on the systematics, distribution and ecology of the ichthyofauna of Meghalaya and their bearing on the fish and fisheries of the state. pp vi +530 Ph.D. Thesis . University of Gauhati (Assam).

Srivastava., CB and Singh, VRP (1964) Inter relationship between standard length and body weight of *Cirrhinus mrigala* (Ham.). *Proc. Nat. Acad.. Sci. India Allahabad.* 34(2):37-41.

Thakur, NK and Das . UK (1975) Length-Weight relationship of *H. fossilis* (Bl) *J. Inland fish Soc. India* 6(1974) : 95-96

Venkateshwarlu, T and Banerjee, V (1971) Length-Weight studies in *Gudusia chapra* (Ham.) *Proc. Nat. Acad. Sci. India.* 41(1) : 43-46.

CHAPTER 30

EFFECT OF BODY WEIGHT AND SEX ON LIVER GLYCOGEN LEVEL OF *HETEROPNEUSTES FOSSILIS* (BLOCH.)

B.P. Akela

University Department of Zoology, L.N. Mithila University, Darbhanga - 846 004.

ABSTRACT

The ceneentretion of glycegen in hepatic cells varies according to age, sex, season and environmental factors. The increase in the weight of body caused significant elevation in the level of hepatic glycogen. There existed a direct, positive and linear correlation between body weight and giycogen content of fishes. The coefficient of correlation was statistically highly significant. The increase in the hepatic glycegen level was due to increased rate of glycogenesis upto maturity. The greater amount of glycogen in the liver cells of males than that of females was due to greater feeding intensity of males than that of females.

Key words: *Body weight, Sex, Liver Glycogen Correlation. H fossilis.*

INTRODUCTION

The need for a knowledge of normal values of biochemical parameters has been stressed by several authors. Reviews en the blood chemistry of various fish species have been published by Blaxhall (1972). Aldrin *et. al.* (1982), Hille (l982), Sinha & Banerjee (1988-1989), Banerjee (1988), Verma *et. al.* (1991), Banerjee & Verma (1992), Ghosh & Banerjee (1993), Banerjee (1995), Nath & Banerjee (1996), Chaudhari & Kulkarni (1996), Shashi & Akela (1996), Yadav & Akela (1997), Damodar *et. al.* (2002), Mukhopadhyay *et. al.* (2002) and Yadav & Akela (2002).

Hematelogical changes within fish species have been reported as due to variations in age (Conroy, 1972), season (Herdig & Hegland,1983), strain (Barnhart,1969) and sexual maturation (Lane,1979). Hematology used as an aid in fish health central necessiates a bank of information on normal ranges where variables of this kind are taken into considerations.

Of the various biochemical constituents of fish, glycogen is known to exhibit the most profound variations. Their levels in the body can be used as indicators of the general well-being of the fish. Despite some attempts to express the condition of fish through the quantitative estimates of glycogen seems to have received little attention on this context. The objective of the present study is to correlate the concentration of glycogen in the flesh with biological conditions of air breething fish. The perusal of available literature revealed that very little work has been done on fishes in relation, to body weight in comparison to other vertebrates to extreme difficulty in obtaining satisfactory sample of blood specially from fishes. Therefore, in the present study attempts have been taken to correlate the effect of age and sex on liver glycogen level of common Indian catfish, *H. fossilis.*

MATERIALS AND METHODS

In the present study the common Indian catfish, *Heteropneustes fossilis,* which is commonly known as "Singhi" was used as experimental subject. Fishes were collected from local ponds and paddy fields of Darbhanga with the help of fisherman. It was also purchased from the local Market. The live fishes were brought to the laboratory in earthen pots and kept in aquarium. They were acclimatised at laboratory condition for a fortnight with proper supply of fish-feed. The fishes were divided into two groups, A & B. The body weight of fishes of A group ranged from 32.9 to 80 gm and that of B group from 85.0 to 132 gms. The muscle glycogen was estimated by the method of Carrol *et. al.* (1956) which is a modified method of Roe (1955).

RESULTS AND DISCUSSION

The liver glycogen level of male H. fossilis were 671.09 ±20.121 mg/gm and 661.086 ± 19.008 mg/gm in females. (Table-30.1). The coefficient of correlation (r) were found to be 0.785 and 0.916 in males and females respectively. The values of r exhibited good correlation between weight and liver glycogen level. The r-value in females was statistically highly significant. In B group of male and female fishes the glycogen content of liver were 807,806 ± 20.743 mg/gm and 796.03 ± 19.539 mg/gm respectively. The r-values in males and females were 0.958 ml 0.960 respectively. The correlation between liver glycogen and body weight was highly significant. There occurred sexual variabilities in liver glycogen content. The glycogen content was less in females than that of males.

Table- 30.1

Showing the effect of body weight and sex on liver glycogen level of *Hateropneustes fossilis* (Blech.)

Group	Sex	Range of body weight (in gm)	Liver glycogen (in mg/gm)	r
A	Male	32.90-79.00	671.09±20.121	0.785
	Female	30.00-80.00	661.086±19.008	0.916
B	Male	86.30-132.00	807.806± 20.743	0.958
	Female	85.00-100.00	796.031±19.539	0.960

Portal glucose at physiological concentrations represents the prime stimulator of hepatic glycegen synthesis (Davidson, 1981). An inverse relationship between glycogen content and the activity of basal and insulin dependent glycogen synthesis from glucose in the intact hepatocyte ex vivo, had pointed to a decreased activity of glycogen synthesase I in the liver of fed animals (Hue *et al.,* 1975). The increase in the concentration of glycogen was due to less activity and greater consumption of food. The consumption of greater feed resulted into much amount of glucose and duo to lesser activity the amount of glucose was not consumed that in turn underwent glycogenesis. The higher amount of liver glycegen in males than that of females might be due to consumption of more amount of food by males than that of females.

REFERENCES

Aldrin, J.F. ; Messager,J. U and Baud in, LF. 1982. La boichemic clnique on aquaeulture, Interet et. perspectives . CNEXO Actes Collog. 141 291-326.

Banerjee, V. 1988, Erythrocyte related blood parameters in *Bufo melanestictus* with reference to sex and body weight. Environ. Ecol. 6(4):802-806.

Banerjee.V. and Verme, G. K.1992. Effect of three pesticides on erythrocyte and leucocyte morphology in *Channa punctatus* (Bloch.). Proc. Natl. Symp. Etnerg. Tr. Anim. Haemstol. H Patna : 33-38.

Banerjee,V. 1995, Effect of pesticides methyl parathion and cyperrnethrin on the air-breathing fish, *Hateropneustes fossil* is (Blech). Environ. Ecol. l4(l): 163-165.

Barnhart. R. A. 1969. Effects of certain variables en haematolcgl characteristics of rainbow trout. Trans. Am. Fish Sec. 3:411-418.

Blaxhall.P. C. 1972. The haematologieal assessment of the health of freshwater fish . J. Flsh.Biol.4: 593-604.

Carrol, N.V.; Longley, R.W. and Roe, J. H. 1956. The determination of glyeogen in liver and muscle by use of enthrone reagent. J. Biol. Chora. 220 J 583-593.

Choydhari, R.D. and Kulkarni, A, B 1998. Variations in the glyco content in the reproductive tract of the terrestrial snail, Ceras-tusmoussonianud duo to monoerotophas intoxication. Proceedings of National Symposium on Eco-Environmental impact and organism response. Siddharth Publications. 20:1-5.

Conroy,D.A. 1972. Studies on the haematology of the Atlantic salmon, *Salma salar* I. Symp. Zool. Sec. Lond. 301 101-127.

Damodar, B.; Mukhopadhyay, H; Neegi, A. K. and Akela, B. P. 2002. Effect of changing state of environmtent on certain biochemical parameters of the fish, *Anabas testudineus.* Environ. Ecol.20(2):318-321.

Davidson, M. B. 1981. Metabolism. 30:279-284.

Ghosh, K. and Banerjee,V. 1993. Effects of deltamethrin on haemopoietic tissue and some blood parameters of *Heterepneustes fossilis*. Freshwater Biol. 4(4): 283-288.

Hardig. J. and Hogland, L.B. 1983. On accuracy in estimating fish blood variables. Comp. Biochem. Physiol. 75 A: 35-40.

Hille, S. 1982, A literature review of the blood chemistry of rainbow trout, *Salme gairdneri* Rich. J. Fish. Biol, 20: 535-569.

Lane, HC. 1979. Progressive changes in haematelogy and tissue water if sexually mature trout, *Salmo gairdneri*. Rich. during the autumn and winter. J. Fish. Biol. 15: 425-436.

Mykhopadhyay, H; Bandyopadhyay, D. ; Neogi, A.K. and Akela, B.P. 206: Hemastological changes in Heteropneustes fossilis (Bloch.) with reference to changes of water quality. Environ. Ecol. 20(1): 152-154.

Nath, R. and Banerjee,V. 1996. Haematological and biochomical changes in blood of *Clarias batrachus* due to sulphates of manganese and chromium. Environ. Ecol. 14(l) :163-165.

Roe, J. K. 1955. The determination of sugar in blood and spinal fluid with anthrene reagent. J. Biol. Chem. 212:335-343.

Shashi, S.B. and Akela B. P. 1996. Determination of maturity and spawning period by gonadosematic index and measurement of mean ova diameter in certain teleosts. Enviren. Ecol. 14(2) : 399-403.

Sinha, P.K. and Banerjee.V 1988. Cyclic changes in leucocyte count in *Bufo melanostictus* (Scheneides). Him, J. Env, Zool. 2 : 63-69.

Sinna, P.K. *a*nd Banerjee,V. 1989. Seasonal variation of thrombocyte counts in Bufo melanostictus. Envirori. Ecol. 7(2): 456- 457.

Verma, G.K.; Singh, H. K.P. and Benerjee,V. 1991. Effects of handling stress on blood parameters in the fish, *Anabas tesludineus* (Bloch.) Envircrw Ecol. 9(3): 703-705.

Yadav, T. and Akela B. P. 2002. Effect of thermal stress on certain bio chemical constituents of Channa punctatus. (Bloch.). Environ Ecol. 20(4): 753-756.

CHAPTER 31

Analytical Data on Fecundity, Gonado Somatic Index & Ova Diameter of a Weed Fish, P. Ticto Found in a Tropical Lake

***Bijay Bhushan Prasad*[*], *Nasi Ahmad*[**] & *Md. Zahid Eqbal*[**]**

**Industrial fish & fishes (UGC sponsored) SNS College Motihari*

**Deptt. of Zoology, AHSA College, Near-Stadium, Madhubani (Bihar)*

ABSTRACT

Fecundity, GSI and OD of the test fish P. ticto was studied and found that total fecundity ranged from 1168-5590 in different length/weight groups. The fecundity per gm. of body wt. was 741-849 and per gm. ovary wt. was 3726-5694. Fecundity has been found to be related to fish length, wt. & ovary. The peak values of GSI in the month of May was found to be 1.46394 in male and 19.11486 in female, which fell sharply in June, i.e. male having 0.96106 and female having 9.32705 which coincided with the mature stage and spawning stage respectively. Ova diameter progression studies exhibited high values of ova diameter i.e. 0.83-0.84 mm in June and July forming peak.

Key words: *Fecundity. GSI, Ova diameter, Puntius ticto. Turkaulia,(Bihar) Lake.*

INTRODUCTION

The reproductive potential of a population is a function of the fecundity of females which also determines the amount of rearing facilities as well as determines the density dependent factors affecting population size. Fecundity has a vital bearing on the selection of breeders for production purposes.

Sexual maturation (development of gonads) is an important prerequisite for the continuation of the race in fishes. It depends on many factors including mainly the availability of food. The amount of food eaten largely determines the fecundity of the year classes, as well as the growth rate and sexual maturity. Besides temperature, photoperiod, internal rhythm and hormone system also influence the maturation of gonads in fishes. Sexual maturation among fishes includes seasonal cycle of gonads, maturation, fecundity and spawning etc. Gonado somatic index (GSI) provides real picture of the seasonal cycle in the gonads.

The present study, therefore, is a step further in the culture of major caips. The knowledge on the biological aspects especially the breeding behaviour of Puntius ticto (Ham.) may be helpful in the eradication of this fish (weed fish), from carp nurseries and stocking ponds, if necessary.

MATERIALS AND METHODS

It was estimated on the basis of mature eggs in the ovary. Ovaries of the mature fishes of known length and weight were dissected out and weighed and then kept in 5% formaldehyde for about a week. Three small pieces (W_1,W_2, and W_3) were cut from each ovary. They were weighed and then immersed in Glison's solution (Jhingran *et. al.*, 1988) for easier separation of ova. The number of ova in each piece was counted and the total number of eggs in the ovary was estimated:

$$F = \frac{wxN_1 + N_2 + N_3}{N_1 + N_2 + N_3}$$

Linear regression (Mahajan, 1984) was employed to find out the correlation of fecundity with fish length and weight as well as ovary weight.

Gonadosomatic index (GSI) was calculated to see the possible pro-gression of maturation and spawning.

$$GSI = \frac{\text{Gonad weight}}{\text{weight of fish}} \times 100 \text{ (after Singh and Srivastava, 1991)}$$

Measurement of ova diameter was done following Jayabalan (1986). It was measured under a compound microscope as suggested by Clark (1934) using an ocular micrometer. About 1000 ova from ovaries were measured. A micrometer division was equal to 0.0148 mm. The slide on which the ova lay was not rotated, consequently, the ova were measured at any diameter that fell in line with the micrometer scale, as suggested by Clark (1925). Ova diameter data of different months were plotted to see the maturation and spawning.

OBSERVATION

The total fecundity ranged from 1168-5590 in different length/ weight groups, the mean ova number was calculated to be 3004+ 46.02579 (SD = 489), being 504 + 6.03239/cm of the fish length (Table-31.1). The fecundity per gram of body weight) and the fecundity per gram of ovary weight ranged from 3726-5694 (Mean ova No. 4437.2 + 74.01 %/ gram ovary weight) (Table-31.1, Fig.-31.1, Fig.-31.2)

Table-31.1

Number of Ova per gram ovary weight and per centimetre body length in *Puntius ticto* (Ham.) (Only larger fishes are taken into account)

Length group (cm)	N	Average length (cm.)	Ovary weight range (g)	Average weight (g)	Ova range	Average no. of Ova	No. of Ova/gm. Ov. Wt.	No. of Ova/cm. length
5.35-5.61	22	5.51964	0.213-0.583	0.37785±0.02147	1168-3482	2276±296	6024	412
5.62-5.88	33	5.77466	0.277-0.762	0.57249±0.09146	1575-3896	2853±272	4783	494
5.89-6.15	38	5.98528	0.231-0.856	0.69852±0.04538	2088-4946	3217±309	4605	537
6.16-6.42	16	6.30457	0.164-1.116	0.82291 ±0.07164	1956-5590	3639±334	4224	577
6.43-6.69	9	6.54782	0.138-0.988	0.78556±0.03475	1872-5218	3485±262	4436	532

Mean of Ova No. = 3094, SD = 489, SEM = 46.02579

Mean of Ova No./Cm., body length = 504, SD = 65.52862, SEM = 6.03239

Mean body length = 6.02939

Figure 31.1

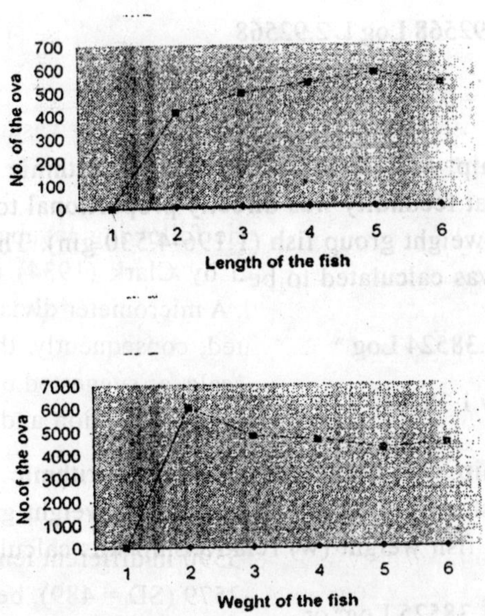

Figure showing no. of ova per length & weight group in Puntius ticto

Figure 31.2

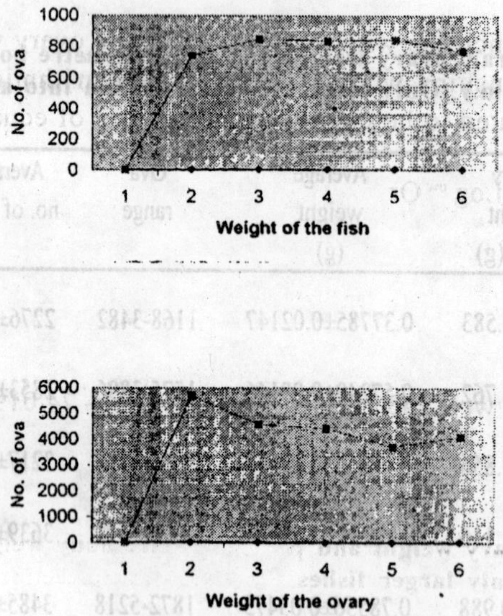

Figure showing no. of ova per body and ovary weight in Puntius ticlo

The relationship between fecundity and length of the fish was determined which showed directly proportional increase in fecundity with increase in length except in the last length group fish (6 43-6.69 cm). The logarithmic transformation of the variables were employed since it gave an approximately linear relationship. The fecundity (F) and the fish length (L) relationship was calculated to be:

Log F - 1.21435+ 2.92568 Log L 2.92568

or F - 16.38136 L (r = 0.15396)

A linear relationship was obtained between logarithmic values of fish weight and fecundity. It was found that fecundity was directly proportional to the weight of the fish with the only exception in last weight group fish (1.196-4.530 gin). The fecundity (F) and the fish weight (W) relationship was calculated to be-

Log F = 2.69649 +1.38524 Log ^{w}

or F = 497.152930 w 1.38524

A linear relationship was obtained between logarithmic values of fish weight and fecundity. It was found fish with the only exception last weight group fish (4.196-4.530 gm) The fecundity (F) and the fish weight (w) relationship was calculated to be-

Log F = 2.69649 + 1.38525 Log^{w} or

F = 497 .15293 w 1.38524

(r = 0.33525)

Mean of ova no./g body wt = 794.8 = 795, SD = 68.43025, SEM = 6.29951 (Table-31.2)

Fecundity was found to increase with the increase in the ovary weight. The fecundity and the ovary weight relationship; when transformed logarithmically, showed linear relationship. It has been calculated to be expressed in the form of equation.

Log F =3.54057+0.56186 Log ^{ow} Or

F =3771.9223 + ow°56186

(r = 0.6943 8)

Mean of ova No./g. ovary wt = 4437.2, SD = 05889, SEM - 74.01962 (Table-31.2)

Table-31.2

Number of Ova per gram ovary weight and per centimetre body weight in Puntius ticto (Ham.) (Only larger fishes are taken into account)

Range of fish weight (g)	N	Average weigth of fish (g)	Ovary weight range (g)	Average weight of ovary (g)	Ova range	Average no. of Ova	No. of Ova/ gm. body wt.	No. of Ova/gm. Ova wt.
2.856-3.190	17	3.056	0.213-0..574	0.39759±0.04934	1168-3482	2264±149	741	5694
3.191-3.525	22	3.451	0.291-0.715	0.63943±0.08372	1526-4775	2931±292	849	4584
3.526-3.860	34	3.782	0.234-0.833	0.71407±0.05592	2130-4946	3168±346	838	4437
3.861-4.195	15	4.089	0.172-1.116	0.92645±0.02381	1844-5590	3452±328	844	3726
4.196-4.530	12	4.294	0.135-0.988	0.80532±0.06447	1756-5268	3316±462	772	4118

Mean of ova No./gm. body weight = 794.8 = 795 ova. SD= 68.43025. SEM = 6.29951

Mean of ova No./gm. ovary weight = 4437.2, SD = 804.05889, SEM = 74.01962

Mean of ovary weight = 0.69657 gm

Mean of over number - 3026.20

The percentage of fishes maturing at each length unit has been shown in Table-31.4. Stage I and II have been treated immature whereas Stage III and onwards including spent fishes as maturing and mature. No. maturing fish was observed in the samples below 40 mm total length. In case of male, 12% were mature at 40 mm, 38% at 45 mm, 50% at 50 mm and 50% or more in the length groups beyond that length. In the case of females 15% were mature at 40 mm. 20% at 45 mm, 40% at 50 mm, 50% at 55 mm and 50% or more in the length groups beyond that length. Since 50% of males and females were mature at 50 mm and 55 mm length respectively, it can be stated that first maturity in this fish was attained at different lengths in both sexes i.e. at 50 mm length in male, whereas at 55 mm length in female. (Fig. 31.4).

Table-31.3

Monthlty fluctuations in Gonodasomatic Index in Male and Female in puntius ticto (Ham.)

Month	N	Mean weight fish (g)	Mean weight Testis (g)	Gondosometic Index (at 0.05 level of confidence.	N	Mean weight of fish (g)	Mean weight of Testis (g)	Gondosometic Index (at 0.05 level of confidence.
Jan	8	2.32248±0.07383	0.02303±0.00826	0.99735±0.06458	12	2.67959±0.29625	0.23537±0.08457	8.71696±1.69372
Feb	11	2.10829±0.14276	0.02335±0.00352	1.10495±0.18509	14	3.17627±0.40318	0.31185±0.04916	9.84962±0.97433
Mar	12	2.62967±0.13250	0.03130±0.00283	1.19037±0.12891	13	3.64935±0.48395	0.44636±0.11371	12.28393±2.22700
Apr	9	2.96393±0.20787	0.03517±0.00130	1.34626±0.09631	14	3.45384±0.21891	0.50549±0.06489	14.65208±1.37469
May	12	3.4354±0.18047	0.05055±0.00922	1.46394±0.19792	17	3.89186±0.60355	0.76234±010157	19.11486±2.14567
Jun	10	3.03294±0.14816	0.02905±0.00366	0.96106±0.09774	15	3.12863±0.42892	0.28984±0.14399	9.32705±1.44562
Jul	12	2.18266±0.06222	0.01863±0.00481	0.84615±0.15316	15	3.44327±0.48308	0.13177±0.04778	3.82419±0.87184
Aug	9	2.55492±0.19458	0.01-846+0.00752	0.72415±0.06137	16	2.93766±0.18684	0.02787±0.01344	0.94928±0.46191
Sep	8	2.76254±0.23018	0.02314±0.00093	0.83593±0.20864	14	3.25495±0.22831	0.06835±0.02129	2.09231±1.23213
Oct	11	3.26915±0.31612	0.03281±0.00633	0.98862±0.19984	15	3.00424±0.44917	0.11846±0.06528	3.91891±1.56636
Nov	12	2.42825±0.21173	0.02743±0.00518	1.12613±0.14653	13	2.79596±0.23208	0.15292±0.07992	544607±1.63637
Dec	10	2.37621±0.18967	0.02404±0.00089	1.01466±0.16382	14	2.83872±0.23771	0.21539±0.06361	7.58419±2.11913
	123				172			

GSI values showed increasing trend from February onwards forming peak in May (GSI = 1.46394 in male, 19.11486 in female) in both sexes. During June GSI values showed sharp fall (GSI-0.96106 in male, 9.32705 in female). The fallen GSI values in June further declined in July and August and thereafter showed improved condition with some fluctuations. The peak in the month of May indicated the mature (ripe) stage whereas the sharp fall in June-July showed the spawning season of the fish). (Table-31.3 & Fig.-31.3).

Figure 31.3

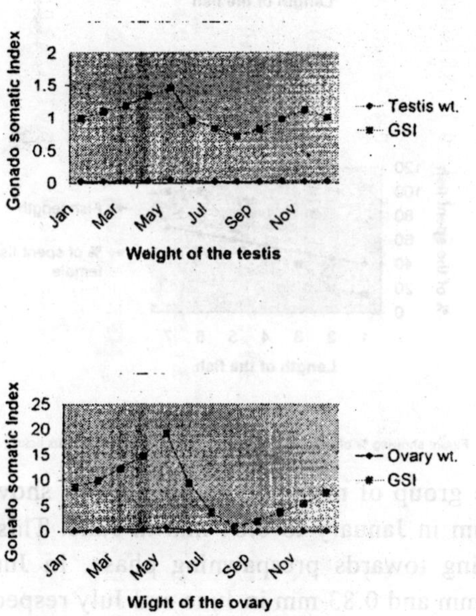

Figure showing Gonado somatic index per wt of testis and ovary in Puntius ticto

Table 31.4

Number and percentage of mature and spent specimens in *Puntius ticto* (Ham.) in each 5 mm length group

Length group (mm)	N	Male No. of mature and spent fish	%age of mature and spent fish	N	Female No. of mature and spent fish	%age of mature and spent fish
38-42	50	6	12.0	20	3	15.0
43-47	13	5	38.4	10	2	20.0
48-52	12	6	50.0	20	8	40.0
53-57	12	9	75.0	13	10	50.0
58-62	10	9	90. 0	13	9	69.2
63-67	12	11	91.6	8	8	100.0
68-72	-	-	-	13	13	100.0

Figure 31.4

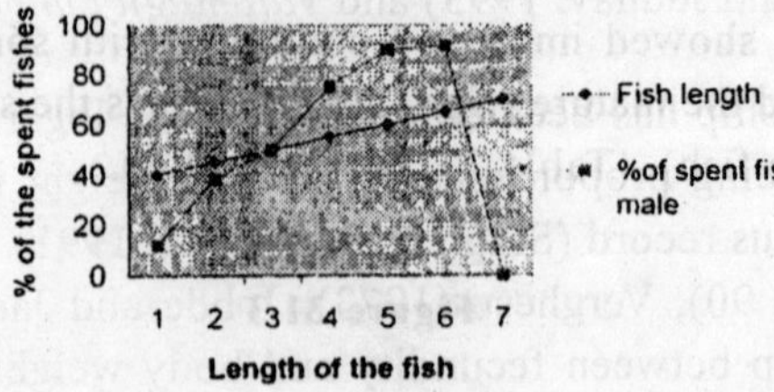

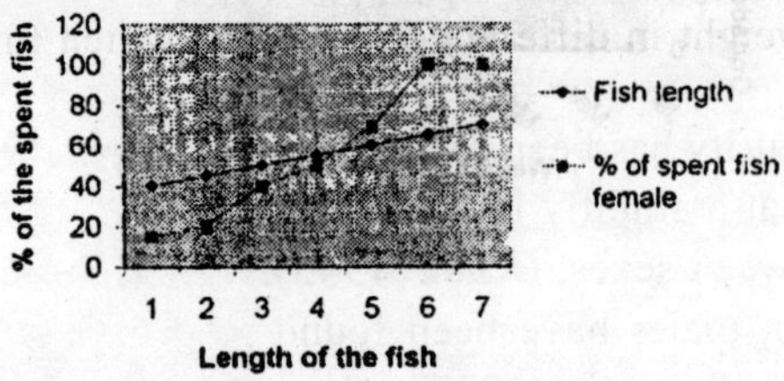

Figure showing % of spent fish in relation to male & female Puntius ticto

There was only one group of maturing oocytes which showed gradual progression in ova diameter from 0.32 mm in January to 0.84 mm in June. This progression indicated the immature stages progressing towards prespawning phase. In June and July ova diameter showed high values (0.84 mm and 0.83 mm in June and July respectively) forming peak. This peak represented the ripe ova of spawning season.

DISCUSSION

Fecundity of *Puntius ticto* ranged from 1168-5590 in different length/weight groups in the fishes of lake Turkaulia (Motihari). The relative fecundity was found to be 504 ova/cm fish length and 795 ova/gin fish weight (with high SD values). These observations, however, slightly differ from the previous informations. Ibrahim (1957) has reported 2368-8590 (average 4570) ova produced, being 990 ova/ gm body weight in the same species from fishry water at Cuttack (Orissa). On the other hand [Choudhuri (1960 a)] has reported 1351-2313 (average 1862) ova in the same species from Cuttack (Orissa). These variations in fecundity values may be due to changed ecological and geographical conditions, as have been reported in *Channa punctata* in which fecundity range has been recorded 2300-296000 from Aligarh (Qasim and Qayyum, 1963), 733-9225 from Bhuvaneslnvar (Rath and Hajmadi, 1976) and 2200 - 33873 from Gunter, Andhra Pradesh (Reddy, 1979a&b).

Fecundity has been found to increase with increasing length/weight but for the larger specimens (6.43 - 6.69 cm) where it decreased slightly. Somewhat similar trend has been found by Ibrahim (1957) in the same species where fecundity was high in specimens 61 - 64 mm long and tended to decrease in smaller as well as larger specimens. There is thus a clear indication that the fecundity of *Puntius ticto* declines after a particular size.

There is linear relationship between the fecundity and fish length. Similar observations have been recorded in *Alosa sapidissmia* (Lehman, 1953), Labeo bata (Bhatnagar, 1972) *Chanda ranga* (Jobde and Jadhav, 1993) and *Valamugil cunnesiue* (Hoda and Qureshi, 1993).

A linear relationship has been obtained between logarithmic values of fish weight and fecundity : fecundity being proportional to the fish weight in most of the cases. This finds support from the previous record (Singh & Srivastava, 1991). Moreover, Reddv (1979 a & b), Reddy & Rao (1989 & 90), Verghese (1973), Jobde and Jadhav (1993) have also recorded straight line relationship between fecundity and body weight in various fish species.

The number of ova was found to increase with increasing weight of the ovary which is in agreement, with the previous informa-tion (Ibrahim, 1957). Rao (1974), Qasim and Qayyum (1961), Das (1964) and Jobde and Jadhav (1993) observed a straight line relationship between fecundity and ovary weight in different fish species which fully support the present findings.

Size at first maturity has been found to be 50 mm in males and 55 mm in females. Male *Puntius ticto* matures at a slightly lower length than the female. This indicates difference in linear growth rate between sexes. (Khan, 1945). Males appear to grow faster than female. In other fish species also, males have been found to mature at a lower length than the female Rao (1974); Pathak & Jhingran (1977). However, reverse is the case of *Mugil cephalus* Rangasawamy (1975).

GSI values increase with the maturation of the fish, become maximum during the peak of maturity and decrease abruptly and sharply when the fish becomes spent Khan (1945); Ganpatti & Chako (1945), Pathak and Jhingran, 1977; Piska & Devi, 1993.

The highest GSI values in May forming a peak indicates the full maturity (ripeness) of the fish. The sharp fall in the index after the peak in May indicates that the fish spawns after May.

In the present study, the mean maximum diameter of ova was recorded to be 0.84 mm, forming the peak. However, average ova diameter 0.70 mm (Rao, 1974) and 0.68 mm (Choudhary, 1960a) have been reported in the same species.

REFERENCES

Bhatnagar, G.K.; 1972: Maturity, Fecundity, spawning season and certain related aspects of *Labeo fembratus* (Bloch) of river Narmada near Hoshangabad. J. Bombay, Nat. Hist. Soc.

Clark, F.N.; 1925: The life history of *Leureshes tennis* an athesine fish with tide controlled spawning habits *Fish. Bull.* 42: 1 -49.

Clark, F.M; 1934: Maturity of the California Sardine (Sardmella caerula) as determined by ova diameter measurements. Fish. Bull. 42: 1-49.

Chaudhary, H. 1960a: Experiments on induced spawning Indian Carps with pituitary injections. *Indian. J. Fish.* 7(1): 20-49.

Das, S.M.; 1964: A study on the fecundity of some freshwater fishes of India with a note on a new concept of comparative fecundity. *Ichthvologica.* 3 - 33 - 36.

Ganpati, S.V. and P.I. Chacko; 1954: Some, observations on the spawning of Indian carps in the 'Bundhs' of Bengal. Indian *Geor. J.* 27 (3 &4) : 1-17.

Ibrahim, K.H. 1957: Bionomics of Forage fishes. Observation on the fecundity of three common species of minor barbels. Jour. Bomb. Nat. Hist. Soc. 54 (4) : 826-834.

Jayabalan, N.; 1986: Reproductive biology of silver belly. *Leiognathus* splendens (cuvier) at Porto Novo. Indian *J. Fish.* 33 (2): 171 -179.

Jlimgran, V.G.; A. V. Natrajan; S.M. Bannerjee and A. David. 1988: Methodology on reservoir fisheries investigations in India. Bulletin No.2, Reprinted May, 1988 CIFRI, Barrackpore (WB).

Jobde, P.V. and P.O. Jadhav; 1993: Study on the fecundity of a teleost fish *Chanda ranga* (Ham. Buch). Biovigyanam, 19 (1 & 2) : 92-95.

Khan, H. 1945: Observations on the spawning behaviour of carp in Punjab. *Proc. Nat. Inst. Sci.* India 11:315 -320.

Mahajan, B.K.; 1984: Methods in biostatistics pp. 304. Printed in India by Leipzig Press, D-52. South Extension Part-1, New Delhi.

Malhotra, Y.R. and Sharma, K; 1979: Studies on Fecundity of *Gara imata* (Ham.) *Mystus vittatus* (Bloch) from Jainmu, 11 All India Seminar on Ichthyology P. 13

Pathak, S.C. and A.G. Jhingran; 1977: Maturity and fecundity of *Labeo calbasu* (Ham.) of Loni reservoir. Madhya Pradesh. J, 12(1) : 60-62.

Piska. Ravi shankar and Rama Devi; 1993: An account of fecundity in the freshwater catfish, *Heterpneusics fossilis* (Bloch) of lower Manair reservoir, Karimnagar, Bio journal, Vol. 5, No. 1 & 2, 127-129.

Qasim, S.Z. and A. Qayyurn: 1961: Spawning frequencies and breeding season of some fresh water fishes with special reference to those occurring in the plains of North India. *Indian J. fish. 8 (1):* 24-43.

Qasim S.Z. and Qayyum, A., 1963: Fecundities of some freshwater fishes. *Proc. Nat. Inst. Sci. India 29* (B): 375-381.

Rangaswamy, C.P.; 1975: Maturity and spawning of *Mugil cephalus* Linnaeus of Lake Pulicat. In. R. Natrajan (Ed.) Recent researches in estuarine biology. Hindustan Publications. Delhi, pp. 47-60.

Rao, G.R.; 1974: Observations on the age and growth. Maturity and fecundity on *Labeo fimbriatus* (Bloch) of the river Godavan. Indian *J. Fish.* 21 (2): 427-444.

Rath, R. and Hajmadi, P.M.; 1976: The fecundity of *opplnccphalus punctatus* (Pisces : Channidae). *J. Zool. Soc.* India. (1&2) : 1-6.

Reddy, B.B.S.; 1979a: Maturity and spawning in the murrel *Channa Punctat*a (Bloch, 1773) (Pisces. Teleostei. Channidae) from Guntur, Andhra Pradesh. *Proc. Indian Nat. Sci. Acad. B.* 45 (6) : 543-553.

Reddy, P.B.S.; 1979b: The fecundity of *Channa punctata* (bloch, 1793) (Pisces- Teleostei, Channidae) From Guntur, India. *Proc. Indian. Acad. Sci. 88 B.* Part 1 (2): 95 - 98.

Reddy, Y. Siva and M. Babu Rao, 1989 & 1990: Spawning and fecundity of *Channa puntata* (Bloch) (Pisces: Channidae) from polluted lake, Hussain Sagar, Hyderabad, India. Matsya. 15 & 16 : 82-87.

Singh, S.R. and V.K. Srivastava; 1991. Observations on the gonadosomatic index and fecundity of the Ganga river prawn *Macrobrachium binnanicum choprai* (Tiwari). *J. Adv. Zool.* 12(1): 50-55.

CHAPTER 32

STUDIES ON THE BIOLOGY OF *PUNTIUS SOPHORE (HAM)* FROM THE LOKTAK LAKE, MANIPUR WITH A NOTE ON ITS UTILIZATION IN THE REGION

O. Shashikumar Singh[*] and H. Tombi Singh[**]

[*]*Fishery Research Laboratory, Post Graduate Dept. of Zoology, D.M. College of Science, Imphal-795001, Manipur, INDIA.*

[**]*Institute of Bioresources and Sustainable Development, Takyelpat, Imphal-795001, Manipur, India.*

ABSTRACT

The biology of the cyprinid weedfish *P. sophore*(Ham) collected from the Loktak Lake, Manipur have been studied with reference to its length-weight relationship and condition. Altogether 309 specimens of the fish of different size and sex collected during the period January 2003 to December 2003 were used in the studies. Results of the length weight studies showed the applicability of the cube law to *P. sophore* (a = 2.675; r = 0.98). Co-efficient of condition, K, for both sexes combined was found to vary from 1.13 to 1.60 with a mean value of 1.38. Its monthly value ranged from 1.27 in April to 1.52 in September while relative condition factor, Kn, also fluctuated from a minimum of 0.96 in July to 1.16 in September.

The utilization of the fish in Manipur for preparation of the fermented fish product "Ngaari" has been discussed.

Key words: ***Weed fish, Length-weight relationship, condition, fermented fish.***

INTRODUCTION

***Puntius sophore (Hamilton)* (Cypriniformes:Cyprinidae) is one of the most common carp minnows found abundantly in ponds, streams, rivers and lakes in Manipur. Earlier studies**

have shown that the so-called trash or weed fishes like *P. sophore* play significant roles in a pond ecosystem mainly by affecting the food chain of the cultured carps (Natarajan and Ramakrishna, 1975). The fish is a bottom and marginal browser competing to a large extent with *Cirrhina mrigala, Labeo calbasu* and *Labeo rohita,* thus adversly affecting their growth and proliferation. Moreover the weed fishes breed usally in the premonsoon months of March and April. Thus, their fry and fingerlings are already abundant in the ponds by the monsoon months when the cultured carps generally breed. These weed fish fries compete with the carp fries for food and space and also prey upon the carp hatchlings leading to a higher mortality of the cultured fishes (Alikunhi, 1952). In addition to this, the weed fishes are found to be dangerous destroyers of carp eggs.

Though regarded as a trash fish in other parts of the country, *P. sophore* is of plausible commercial importance and consumer acceptability in Manipur. Chaudhury and Banerjee (1965) had reported that the annual catch of *P. sophore* from various water bodies of Manipur including the Loktak Lake represented upto about 40% of the total fish catch in the State.

P. sophore is consumed in various ways such as in fresh, sun-dried, smoked or fermented. The special use of the fish is in the preparation of the indigenous fermented product locally called as "Ngaari" or "Utong-nga", which forms an essential daily culinary item for a typical Manipuri family.

The annual demand of *P. sophore* for extensive consumption in sundried, smoked, semi-fermented or fermented form is to the tune of 555 metric tonnes (Chaudhury and Banerjee, 1965) most of which is fulfilled through import from outside the State. This calls for manifold increase in the production of the fish through capture as well as culture fishery in Manipur.

Despite its importance as a fish of plausible commercial importance, there are no detailed report on the biology of P. *sophore* except that of Chaudhury & Banerjea, (1965) and Natarajan *et al* (1975). The only report on the trash fishes found in Manipur is that of Singh (1985) who made certain investigations on the biology of *Amblypharyngodon mola (Ham)* collected from Kharungpat Lake, Manipur. Hence, the present study on the biology *of Puntius sophore (Ham)* had been taken up.

MATERIALS AND METHODS

309 specimens of *P. sophore* were collected for 12 months from January '03 to December '03 from different landing sites around the Loktak Lake (24°25' - 24°40'N) latitudes and 93°45' - 93°55'E) in the Bishnupur District, Manipur. The specimens were preserved in 10% formalin. Measurement of length, weight and other parameters were done in the Fishery Research Laboratory of the D.M. College of Science, Imphal.

'Total length' was the length of the fish from the tip of the snout with the jaws closed to the end of the longest caudal fin ray. 'Standard length' was the length measured from the tip of the snout with the jaws closed to the caudal peduncle where the vertebral column apparently ends. 'Weight' was the total weight of the fish with all organs and structures intact. Lengths were measured to the nearest 0.01cm using a fish measuring board with attached meter scales. Weights were measured to the nearest 0.01g using a single pan electric balance after blotting the specimens dry with blotting paper.

The relationship between length and weight was analysed using the general formula given by Debraj (1980):

$W = a L^b$, where L = total length in mm,

W = total weight in g,

a & b are constants.

The actual values of the constants 'a', referred to as the initial growth index and the exponent 'b'. variously called the 'equillibrium constant(Sinha, 1972) or the regression slope(Vinci and Kesavan Nair, 1974) were calculated using the formula of Lagler (1952).Condition factor, K and Relative condition factor, Kn were calculated by the follwing formulae.

$$\text{Condition factor, } K = \frac{W \times 10^5}{L^3}$$

where W = weight in g, L = total length in cm. 10^5 was a factor used to bring the value of K near unity.

$$\text{Relative condition factor, } Kn = \frac{W}{\overline{W}}$$

Where W and $\overline{W}$ are the observed and calculated weights, respectively.

Information on the procurement of *P. sophore* and preparation of fermented product 'Ngaari' were collected from the local fish importers and fermenters

RESULTS AND DISCUSSION

Length-weight relationship

The length-weight relationship of *P. sophore,* based on 309 specimens ranging from 52. - 94 mm in total length and 2.31 - 11.25 g in weight (Table 32.1), were studied for both sexes combined. The general equation depicting the mathematical relation between the two parameters was calculated as follows.

$W = 0.000056156\, L^{268}$ (r = 0.98)

Where, W and L are the total weight in g and the total length in mm respectively.

The logarithmic relationship between the two parameters was as follows:

Log W = - 4.2507 + 2.675 log L.

Table 32.1

Length-weight data of *Pimthis sophore* from Loktak Lake, Manipur.

Sl.No.	Length Class (mm)	Mean Length (mm)	Range in weight (8)	Mean weigh (8)	No. of fish
1.	50-54	50.0	1.38- 1.52	1.45	12
2.	55-59	57.6	2.6 -2.9	2.66	19
3.	60-64	65.7	2.38 -4.54	3.33	63
4.	65 -69	66.6	3.10-4.70	4.00	74
5.	70-74	71.7	4.28 - 6.40	5.01	39
6.	75 - 79	76.9	5.40-6.78	6.27	42
7.	80- 84	80.0	6.74-6.82	6.78	10
8.	85-89	87.0	8.70 - 9.05	8.85	30
9.	90-94	92.0	90- 10.5	9.90	20

The scatter diagram (Fig. 32.1) drawn using the untreated individual data showed a parabolic relation between the length and weight indicating the applicability of the general formula W = aLb to *P. sophore.* However, a straight line relationship was obtained when logarithmic values of the length and weight were plotted (Fig. 32.2).

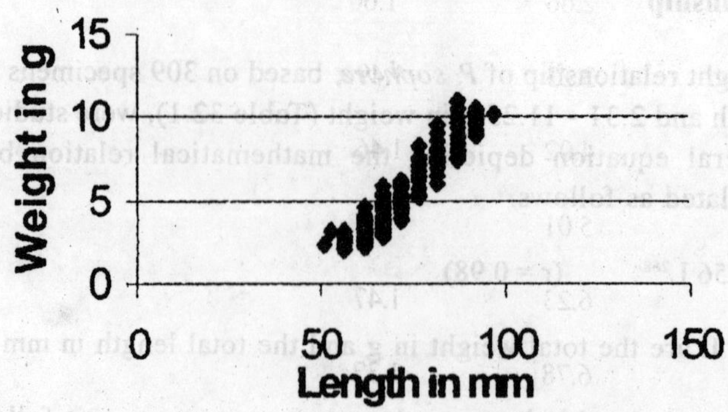

32.1 Length-Weight scatter in P. sophore

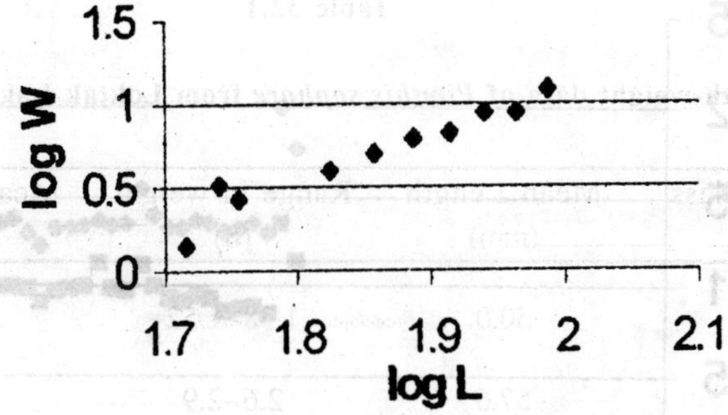

Fig. 32.2 Scatter diagram log L & log W

Condition factor, K

The value of the co-efficient of condition (K) for both sexes combined was found to vary from 1.13 to 1.60 with a mean value of 1.38 (Table 32.2). The value did not show wide fluctuations with varying lengths (Fig. 32.3). Howeve, two distinct peaks were observed at the sizes 54 mm and 74 mm. The peaks at 54 mm and 74 mm are followed by sudden falls in the K value at the sizes 56 mm and 76 mm, respectively. These sudden falls in the K values may be an indication of the sizes at which the fish spawned. Another inflection was also observed at the size 88 mm which may be the consequence of another spawning at this size.

Table 32.2

Condition Factor(K) and Relative Condition Factor (Kn) of *Puntius sophore* from the Loktak Lake, Manipur.

Sl. No.	TL (mm)	W(g)	K	Mean	Kn	Mean
1.	50.0	1.45	1.16		0.77	
2.	55.0	2.66	1.60		1.09	
3.	60.0	3.23	1.495		1.01	
4.	65.0	4.02	1.463		1.03	
5.	70.0	5.01	1.13	1.38	0.98	1.02
6.	75.0	6.23	1.476		1.00	
7.	80.0	6.78	1.324		1.26	
8.	85.0	8.85	1.441		1.02	
9.	90.0	9.90	1.358		1.00	

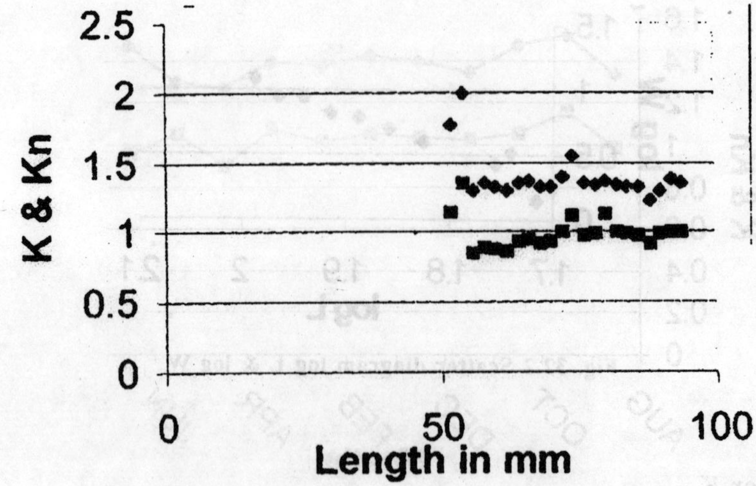

Fig. 32.3 Size variation on K & Kn

Monthly variations in K

The values of K showed monthly fluctuations with a minimum value of 1.27 in April to a maximum of 1.52 in September. When the monthly K values were plotted in a graph a distinct peak was obtained in the month of September (Fig. 32.4). Two inflections, one in November and another in April were also observed. According to Hart (1946), these inflections in the K value can be taken as fair indication of the spawning months of the fish. At the same time, 50% of the fishes collected in the month of April were found to have fully matured gonads. However, none of the fishes collected in November had matured gonads. The fall in the K value of a fish in the non-breeding months may be due to low food intake, adverse environmental conditions and high degree of parasitism (Papageorgiu, 1979).Thus, it may be concluded that *P. sophore* breeds once in a year during the months of March, April and May. This is further substantiated by the observa-tion of small fishes in the months of July and August.

Relative condition factor, Kn

The use of condition factor(K) in interpreting the general well-being of a fish has many drawbacks. The K value is affected by length as well as several other factors like environment, supply of food, degree of parasitism, and above all, the sexual cycle of the fish (Papageorgiu,1979). To remove the factors adversely affecting the correct interpretation of the general well-being and adaptiveness of the fish, another factor, the "relative condition factor(Kn)" have been used by many workers (Narasimha, 1970; Hanumantharao, 1974).

The value of Kn was found to vary from a minimum of 0.77 to a maximum of 1.265 (Table 32.2). The Kn value showed no regular increase or decrease with length. However, two distinct peaks were observed, one at size 74 mm and another at 80 mm, which more or less coincided with the sizes at maturity and spawning of the fish. This showed that the general condition of the fish is high during the prespawning and first spawning sizes of the fish. This further explained the high survival of the fish leading to active proliferation of the fish in the waters of Manipur.

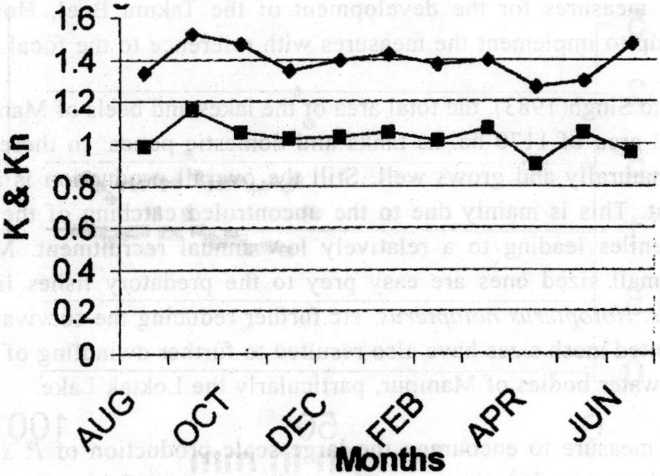

Fig. 32.4 Monthly variation in K & Kn

Monthly variation in Kn

Seasonal values of relative condition factor, Kn, were found fluctuating within the range 0.90 -1,16, the minimum being observed in April and the maximum in September. . From these observations, it becomes evident that the breeding cycle have a significant effect on the condition of the fish and this effect cannot be removed even with the use of the relative condition factor. The low values of Kn, thus coincides with the spawning months of the fish, i.e. from April to June.

From the study it can be concluded that *P. sophore* is a fish with considerable ecological importance in the Loktak Lake of Manipur. The breeding season of the fish is also well ahead of the major carps. Thus, in a natural water body, *P. sophore* may be of great danger to the eggs and spawn of other fishes such as the major carps. However, despite its ecological role as an unwanted weedfish which competes with the commercially more important fish species, *P.sophore* is also a fish with significant commercial importance in Manipur. Apart from its consumption in fresh, dried or smoked condition, the fish is specially used in preparation of the fermented product 'Ngari' or 'Ootonga' which forms an essential culinary item for every typical Manipuri Meitei family from the poorest to the richest. Singh(1984) reported that the demand of the fermented product is much higher than its production in the State, mainly due to the want of the raw material. The annual requirement of 555 metric tonnes approximately of semi-fermented *P. sophore* for preparation of the fermented product 'Ngari' is imported from Assam as well as from Bangladesh resulting to an economic loss to the State of Manipur as well as to the the country. With a view to encourage largescale production of the fermented fish in the State, the Government of Manipur drew up a plan in 1965 to develop the Takmu area of the Loktak Lake for production of *P.sophore* and to meet the demand of 'Phabou', the semi-fermented raw material for preparartion of 'Ngari'. Subsequently, the State Fisheries Department took up a scheme entitled "Dry Fish Production Scheme in Manipur", with an estimated capital investment of 2.60 lakhs. Accordingly a brief survey of the fishery resources and developmental potentials of the place with special reference to large scale production of *P. sophore* (Chatterji and Banerjea (1965). They suggested certain measures for the development of the Takmu Beel. However, no proper steps were taken up to implement the measures with reference to the focal aim.

According to Singh(1983), the total area of the lakes and beels of Manipur is 23,268 ha. with an additional area of 1170 ha. as tanks and domestic ponds. In these vast water areas *P.ophore* breeds naturally and grows well. Still the overall production is always below the actual requirement. This is mainly due to the uncontroled catching of the fish of all sizes, specially the juveniles leading to a relatively low annual recruitment. Morever, the fish, particularly the small sized ones are easy prey to the predatory fishes like *Wallago attu, Channa stheicatus, Notopterus notopterus,* etc.further reducing the survival rate. The use of gill nets of prohibited mesh sizes have also resulted to further dwindling of the population of *P. sophore* in the water bodies of Manipur, particularly the Loktak Lake.

Thus, as a measure to encourage the large scale production of *P. sophore,* the local fishermen should be persuaded not to capture small, immature fishes. Another measure is to enforce the existing legislation of prohibiting the capture of gravid fishes during the breeding season. If such steps are taken up seriously the production of *P. sophore* from the Loktak Lake and other water bodies of Manipur will be enough to meet the annual demand of the fish for preparation of the fermented product 'Ngari'.

REFERENCES

Alikunhi, K.H. 1952. On the food of young carp fry J. Zool. Soc. India. 4 : 77 - 84.

Chaudhary and Banerjee, S. M. 1965. Report on the fisheries of Manipur with reference to the developing of Takmu Beel area of the Loktak Lake. CIFR1 BULL. Misc. Contribution No. 4 Aug 1965.

Hanumantharao, L.1974. Studies on the biology of *C. mrigala* (Ham) of the R. Godavari. Indian J. Fish. 21(2) : 303 -310.

Lagler, K.F. 1952 (Ed). Ichthyology. John Wiley and Sons. New York Santa Barbara London Sydny Toronto pp. 506 + xv.

Natarajan, A. V, Ramakrishna, M. and Khan, M.A. 1975. The food spectrum of trash fishes in relation to major carps in Konar and Tilaiya Reservoirs (Bihar). J. Inland Fish. Soc. India. 7 : 65 - 75.

Narasimhan, K.A. 1970. On the length-weight relationship and relative condition in *Trichiunts lepturus* Linn. Indian J. Fish. 17 (1 & 2) : 90 - 98.

Papageorgiu, N.K. 1979. The length-weight relationship, age, growth and reproduction of the roach, *Rutilus rutilus (L)* in the Lake Volvi. J. Fish. Biol. 14 : 529 - 536.

Singh, H.T. 1983. On the occurrence and bionomics of Osteobrama belangeri in Manipur Valley. Proc. 70th Session Indian Sc. Cong.,Tirupati, 1983.

Singh, O.S.1985. Length-weight relationship and condition factor in the common fresh water weedfish *Amblypharyngodon mola* (Ham). Oriental Zoologist. 4:13- 18.

Vinci, G.K. and Kesavan Nair, A.K.I974. Length-weight relationship in the threadfin - bream *Nemipterus japonicus* along the Kerala Coast. Indian J. Fish. 21(1) : 299 -302.

CHAPTER 33

Heavy Metal Toxicity Induced Behavioural and Morphological Changes in an Air-Breathing Fish *Channa punctata* (Bloch)

Manish Chandra Varma and Pawan Kumar Thakur*
University Department of Zoology T.M. Bhagalpur University BHAGALPUR - 812007
**P.G.Dept. of Zoology, Rajendra College, Chapra*

ABSTRACT

Renewal Bioassay Tests have been used to evaluate the lethal effects of $HgCl_2$ and $CuSO_4 . 5H_2O$. The 96 - h LC_{50} values for Hg and Cu were found to be 0.76 ppm and 2.31 ppm respectively. Further, the effects of exposure of sub-lethal doses of Hg and Cu on the behavior and morphological changes of the fish have been studied. The toxicity data showed that mercury is more toxic than copper to freshwater murrel, *Channa punctata (Block).*

Key Words: Heavy Metal Toxicity, LC_{50}, Behavioural and Morphological changes.

INTRODUCTION

A considerable number of contaminants of environment contain heavy metals as their constituents (Atchinson *et al,* 1987). The term "heavy metal" generally refers to those metals having density greater than five (Varma, 1990). Mercury is considered to be one of the most toxic metals (Varma, 1998). Considerable information are available regarding the toxicity of mercuric chloride and other heavy metals (Davis *et al* 1973; Eaton, 1973; Brooks *et al* 1976; Varma *et al* 1991, Singh, 2003).

Industries discharge heavy metals recklessly into the environment The consumption of heavy metals particularly Mercury through food chain via fish by human beings cause severe disorders in the physiology. Mercuric chloride induced various harmful changes in animals have been studied. Copper are another toxic heavy metal causing serious damages to entire ecosystem including fishes (Sastry and Sharma, 1980).

The present study is aimed to evaluate the comparative toxicity of Mercury and Copper on a freshwater teleost, *Channa punctata.* Attempts have been made to establish the lethal dose of both the metals, the induced behavioral changes and allied morphological changes in the animals.

MATERIALS AND METHODS

Live specimens of *Channa punctata* of body weight 34 - 36 gm were acclimatized for 15 days in the laboratory. Renewal bioassay tests were carried out for 96 hours with 20 litres of water in plexiglass aquarium. The aquarium water was changed daily restoring the mercury and copper level at each change. The physico-chemical characteristics was analyzed as per methods of APHA, 1976. The ecological conditions of water was maintained as temperature was kept in the range of 27 - 30° C, pH 7.1 - 7.4, DO_2 6.8 - 7.0 ppm, alkalinity 120-126 ppm.

Dissolving appropriate amount of made 1000 ppm of stock solution $CuSO_4$. $5H_2O$ and HgC12 in deionized water.

A group of fish was exposed to each concentration and mortality was noted after 24 h, 48 h, 72 h, and 96 h. The percent mortality and probable mortality were plotted on graph against log concentration. The LC_{50} values were calculated by both graphical interpolation and Probit analysis method (Finney, 1964).

RESULTS AND DISCUSSIONS

Exposure of fishes to different concentrations of mercury showed following trend:

No mortality up to 0.4 ppm, 20% mortality at 0.6 ppm and 100% mortality at 1.2 ppm (Table - 33.1).

Table - 33.1: Mortality of *Channa punctata* at different concentrations of Hg after 96 hours exposure.

Cones. (ppm)	Log conc.	No. Of fishes exposed	No. of fishes dead	Percent Mortality	Probit Kill
0.2	0.700	10	-	-	-
0.4	-0.398	10	-	-	1
0.6	-0.222	10	2	20%	4.15
0.8	-0.969	10	6	60%	5.25
1.0	0.000	10	10	100%	8.72

Exposure of fishes at different concentrations of copper showed following trend;

No mortality up to 1.0 ppm, 20% mortality at 1.5 ppm and 100% mortality at 3.5 ppm (Table -33.2)

Table- 33.2: Mortality of *Chana punctatu* at different concentrations of cu after 96 hours exposure.

Concs. (ppm)	Log conc.	No Of fishes exposed	No of fishes dead	Percent Mortality	Probit Kill
1.0	0.000	10	-	-	
1.5	0.176	10	2	20%	I4.16
2.0	0.301	10	4	40%	4.74
2.5	0.398	10	7	70%	5.52
3.0	0.477	10	8	80%	5.84
3.5	0.544	10	10	100%	8.72

When percent mortalities of fishes were plotted against different log concentrations of mercury and copper, the typical sigmoid curves were obtained for both the metals. This plot gave the LC_{50} values as 0.78 ppm and 2.13 ppm for Hg and Cu respectively. When probit kill, calculated from percent mortality was plotted against respective log concentrations, straight lines were obtained. LC_{50} values were calculated after this graphical plot was 0.75 ppm and 2.13 ppm respectively for Hg and Cu. In case of Hg, the average LC_{50} value after employing both the graphical methods was 0.76 ppm. Dhanekar *et al.,* 1985 reported 96 h LC_{50} value as 1.0 ppm of mercuric chloride to *Channa punctata.* The slight difference in the result obtained in present study might be due to the size, age and sex of the fish.

CHANGE IN THE BEHAVIOR OF FISH AFTER EXPOSURE TO HG AND CU

On exposure to Hg

Loss of balance, discoloration, opaque eyes curved spine and stiffness in the body was observed. Increase in the opercular beat with the increase in the concentrations was marked. Increased opecular movement with decreased swimming was also observed by Dhanekar *et al,* 1985 in $HgCL_2$ exposed fishes.

On exposure to Cu

Jerky movements of the fishes were marked. Loss of balance and excessive mucus secretion was noticed. In some dead fishes, reddish spots were present at the base of dorsal, pectoral and caudal fins. The opecular movement rate increased with the increase in exposure period and concentrations.

The gills are known to accumulate significant amount of heavy metals (Shaw & Brown. 1974; Wobesser, 1975) which causes destruction of respiratory surface and coagulation of mucus on gills.

Hg was more toxic to *Channa punctata* than Cu. Cairns *et al* (1975) also reported Hg Cl_2 to be far more toxic than all others expect $AgNO_3$.

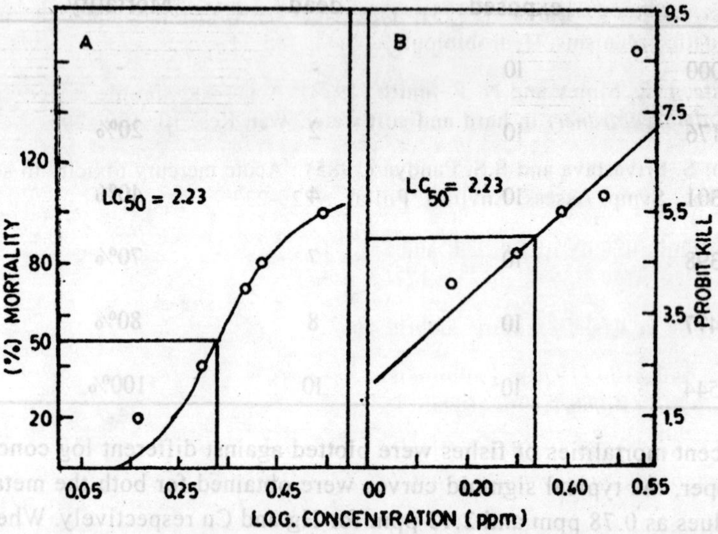

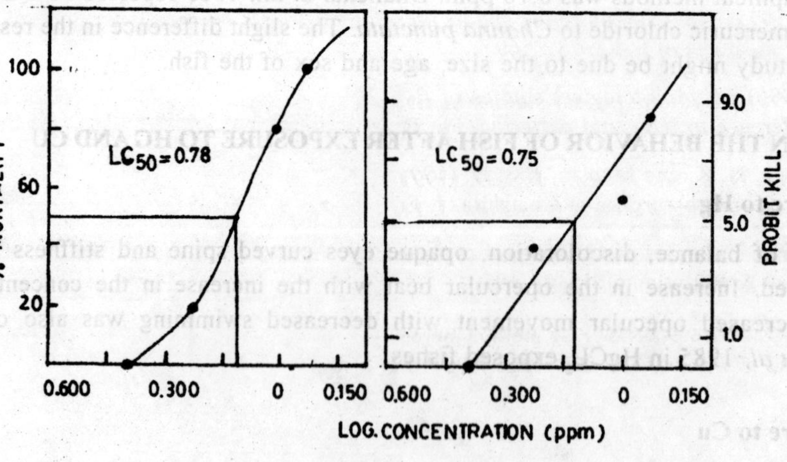

REFERENCES

APHA (1976): Standard method for the examination of water and wastewater. 14th edition, Am. Pub. Health. Assoc., Washington, USA

Atchinson, G. J., Henry Mary, G and Mark B. Sandheinrich (1978): Effects of metal on fish behavior. A review. Environmental Biology of fishes 18(1): 11-25

Brooks, R P; J. R. Lewis and R. D. Reeves (1976): Mercury and other heavy metals in trout of Central North Island, New Zealand, N. Z. J. Mar. Freshwater, 10: 223-224.

Cairns, J; A. G. Health and B. C. Barker (1975): The effect of temperature upon the toxicity of chemicals to aquatic organisms. Hydrobiologia, 47 (1): 135-71.

Davis, P. H.; J. P. Goette; I. R. Sinley and N. F. Smith (1973): Acute and chronic toxicity of lead to rainbow trout *Salmon gairdneri* in hard and soft water. Wat. Res. 10: 199 -206.

Dhanekar, S; K. S. Rao; S. Srivastava and S.S. Pandya (1985) : Acute mercury toxicity to some fresh water fishes . Proc. Symp. Assess. Environ. Pollut. , 229-233.

Eaton, J. C. (1973): Chronic toxicity of Cu, Cd, and Zn mixture to the fathead minnow *(Pinephales pronelas* Refineswue) Wat. Res. 7: 1723 - 1736.

Finney, D. J. (1964); Probit analysis, 2nd ed. (Cambridge Univ. Press) p. 20.

Gupta, A. K. and U&.K. Rajbanshi (1979) : Pathological changes resulting from bioassay of copper to *Heteropneustes fossils* (Bloch.). Proc. Symp. Environ. Biol. , 167- 172.

Sastry , K. V. and K. Sharma (1980): Effect of $HgCl_2$ on the activities of brain enzymes in a fresh water teleost, *Ophiocephalus (Channa) punctata* Arch. Environ. Contam. Toxicol 9 : 425 - 430.

Shaw, T. L. and V. M. Brown (1974) : The toxicity of some forms of copper to rainbow trout Water. Res., 8 : 377-382.

Singh, S. K. (2003) : Ecological study of water body of South Bihar, Ph. D. Thesis, T. M. Bhagalpur University, Bhagalpur

Thangappam, R. (1972) : Mercury pollution and its preservation with particular reference to chloralkali plant. Chemical age of India. 23 : 569.

Varma, M. C. (1990) : Hydrobiological studies of river Subernarekha at Ghatsila (Singhbhum) with special reference to industrial pollution. Ph. D. Thesis, OBhagalpur University, Bhagalpur

Varma, M. C. Singh, N. K. and Munshi, J. S. D. (1991) : Algal species diversity as an index of water quality of river Subernrekha at Ghatshila. J. Freshwater. Bio. 3 : 235 - 239

Varma, M.C. (1998): Chemical and Biological evaluation of an industrially polluted river. Journal of environment & Pollution 5(2): 181-107

Wobesser, G. (1975) : Acute toxicity of methyl mercury chloride and mercury chloride for rainbow trout *(Salmo gairdneri)* Fry and Fingerling J. Fish Res. Bd. Can 32:2005-2014.

CHAPTER 34

STATUS OF FISHERIES RESOURCES IN SELECTED BACKWATERS OF KERALA

P.K. Sukumaran and V.K.Unnithan

Central Inland Fisheries Centre of CICFRI, Alappuzha-688 001, Kerala

ABSTRACT

Seasonal investigations on fishery resources were conducted in ten backwaters of Kerala during 1996-97. The study indicated that the backwaters are under deterioration and the fishery is being lost due to a variety of anthropogenic reasons. Reclamation of the backwaters coupled with the alarmingly low depth ($\bar{x}$: 2.05 m) for nine out of ten systems have reduced the water holding capacity of the backwaters to a large extent. Retting of coconut husk in the backwaters has been the most predominant activity that contributed to the organic pollution resulting in low water quality and depletion of faunal resources. The total landings from different backwaters varied from 96.8 to 2898.7 t. The study reveals that the fishery suffers from both intensive fishing and unscrupulous fishing of undersized fishes.

Keywords: Backwater, lagoons, anthropogenic, reclamation, exploitation, deterioration

INTRODUCTION

The brackish water environment is a unique ecosystem of high fertility, supporting natural fishery in the rural economy of India (Thampi, 1973). Coastal wetlands are a permanent habitat for a variety of resident and migratory aquatic fauna. The Kerala backwaters are interconnected with lagoons, bays and swamps. These backwaters (area: 2,42,600 ha) harbouring about 200 resident, migratory fish and shellfish species. The fishing activities in these backwaters support livelihood about 0.2 million fishermen and provide full-time employment to about 50,000 fishermen. The annual catch from these water bodies spread in nine coastal districts is reported to be about 33,000 t (George & Sebastian, 1970). These water bodies are served as a nursery ground for prawns *viz., Penaeus indicus, P. monodon, Metapenaeus dobsonii, M. monoceros, crabs* and finfishes like mullets, perches and pearlspot and the breeding ground for freshwater caridian prawns *(Macrobrachium* sp.). Obviously, the lake ecosystem has a high carrying capacity and playing a crucial role in both biologically and economically. The need to protect, preserve such ecosystem is imperative. Of late, these ecosystems are reported to be under severe environmental stress arising from a variety of anthropogenic activities. Very little information is available on the backwaters (Shetty, 1965, Nair *et al.*, 1983, Kurup *et al.*, 1993). Scientific monitoring of these ecotypes aiming at its rational fishery management has been highly inadequate. Hence detailed investigations in these water bodies are envisaged to know the present status of the fishery resources.

MATERIAL AND METHODS

Fishery survey was conducted quarterly in ten of the thirty backwaters along the southwest coast of India (Fig. 34.1). The backwaters investigated have a water spread area of 15102 ha, of which 22 % of the total backwaters available in the state. Forty-six landing centres around these backwaters examined for a period of 2-3 days in each quarter either continuously or intermittently in each quarter and the total landings were estimated based on average values for a day. Average fishing days in a month for each category gear was ascertained through repeated enguiry at each stations and the total landings were computed through extrapolation. Fish samples were collected from the major landing centres (Table 34.1). The fishes were identified with the help of Day (1967) and Thalwar and Jhingran (1991).

Table 34.1 List of backwaters and their sea (ha)

Sl.No	Name of the backwater	District	Area (ha)
1	Kadinamkulam (KDK)	Trivandrum	374
2	Anchuthengu (ANG)	Trivandrum	552
3	Ashtamudi (AST)	Kollam	6424
4	Kayamkulam (KYK)	Kollam/Alappuzha	1652
5	Azhikode (AZK)	Trissur	696
6	Chettuva (CTV)	Trissur	714
7	Ponnani (PNI)	Trissur	908
8	Mahe (MHE)	Kannur	88
9	Valapattanam (VPM)	Kannur	3074
10	Neeleswaram (NEL)	Kasargod	825

RESULTS AND DISCUSSION

In Kerala, the backwaters support a rich fishery. Their relative abundance was recorded for the various systems explored. The common species that regularly contributing to the fishery is listed in Table 34.2. Of the 94 species listed, different workers, thereby establishing a close relationship of the backwater fishery with that of marine, have recorded 63 from the

Table 34.2

Total landings (t), gear and fishermen at different backwaters during 1996-97

Backwaters	KDK	ANG	AST	KYK	AZK	CTV	PNI	MHE	VPM	NEL	Average	Average
Area (ha)	374	552	6424	1652	696	714	908	88	3074	825	($kg ha^{-1} yr^{-1}$)	(%)
Major species/groups												
FISHES												
Acanthurus spp.	0.7	0.1	3	3.9	5	5.3	2.9	4.8	8.9	10.5	2.9	0.5
Genrres spp.	7.9	5.5	30.2	29.8	65.3	18.3	12.1	2.7	36.8	54.4	17.1	2.7
Platycephalus indicus	0.6	0.2	32.3	17.1	7	6.4	8.4	1.3	3	27.8	6.8	1.1
Lieognathus spp.	1.2	2.6	9.6	29.8	6.7	12.4	12	5.7	6	18.7	6.8	1.1
Etroplus spp	174	297	168	128.6	38	27	30.3	8.9	357	65.8	35.8	5.7
Megalops cyprinoides	6.1	12.4	11.8	15.9	1.5	0	2.1	3.6	2.7	1.7	3.8	0.6
Arius spp	6.7	22.2	83.8	45.6	37.3	18.6	20.6	7.7	24	42.9	20.2	3.2
Ambassis spp.	13.8	5.4	75.4	14.1	12.5	21.9	8.6	0	5.2	8.2	19	3
O.mossambicus	11.7	31.5	0.7	2.1	15.4	0.1	8.2	0.4	0.9	1.4	5.9	0.9
Stolephorus spp.	0	0	44	7.4	55.5	12	18.1	1.2	6.7	7.2	9.9	1.6
Sillago sihama	2	0.1	11	7.3	63	69	136	37	9.1	18.5	5.1	08
Caranx spp.	8.1	1.4	19.7	20.3	10.4	5.4	3.7	2.9	7.9	45.9	8.2	1.3
Lutjanus spp.	3	2.4	27	17.7	11.2	16.9	29.3	11.2	7.1	18.8	9.4	1.5
Mullets	31.1	26.7	113.4	89.9	95.9	36.5	30.8	1.6	22.9	41.7	32	5.1
Flat fishes	6.4	0.4	3.9	59	2.3	9.2	20.1	0.6	4.8	10.4	7.6	1.2
Half beeks	0.5	0	1.3	0.9	0.5	12.9	4.7	0	2.6	1.4	1.6	0.3
Others	24.4	13.7	271.4	206.6	143.6	41.8	110.3	16.3	56.9	145.8	67.2	10.7
FISH TOTAL	141.6	154.3	906.3	841.7	514.3	251.5	335.7	72.5	240.7	521.2	259.2	41.2
PRAWNS												
Metapenaeus dobsonii	18.8	30	12.55	453.4	1021	172.4	60.7	9	154	214.7	220.9	35.1
M.monoceros	1.9	4.3	238.4	98	146	22.3	19.4	2.8	58	40.7	41.2	6.5
Penaeus indicus	4.7	14.9	210.6	119.7	179	37.6	29.1	8.7	224	60.1	57.9	9.2
P. monodon	0.4	0	24.3	38.7	2	1.2	49.6	0	1.5	0.2	7.7	1.2
Other penaeids	0	0	12.2	15	0.1	2.5	15.9	0	0.6	15.6	4	0.6
Non-penaeids	0.4	0.9	14.9	16.7	0.2	2.5	7.1	0	0.8	0.5	2.9	0.5
PRAWN TOTAL	26.2	50.1	1755.4	741.4	1348.2	238.6	181.7	20.5	438.9	331.9	334.6	53.1
Crabs	20.7	10	237	64.8	49.6	20.5	33.5	5	77.4	38	36.2	5.8
GRAND TOTAL	188.5	214.4	2898.7	1648	1912.1	510.3	550.6	96.8	757	891	630.1	100
CPUE ($kg\ unit^{-1} day^{-1}$)	5.9	3.7	7.2	8.1	13.5	5.4	4.9'	4.6	7.1	10.8		
Yield ($kg\ ha^{-1}$)	543.6	410.7	451.2	997.5	2747.3	714.7	606.4	537.8	246.3	1080		
No. of gears km^{-2}	65	120	52	83	158	100	83	85	51	97		
No. of fishermen km^{-2}	104	145	58	137	152	84	89	99	39	68		

marine waters. In Ashtamudi backwater alone contributed 97 species belonging to 39 families (Nair *et al.*, 1983). Backwater fauna is generally composed of marine and freshwater organisms, which can adapt to waters of different and varying salinities and truly resident estuarine species (Jhingran and Gopalakrishnan, 1971). Nineteen of them used to be predominantly from the rivers/reservoirs are migrated to backwater during monsoon seasons. Definite zonation can be possible due to the distribution of these species *viz., Puntius filamentosus, P. sarana, Labeo dussumieri, Mystus malabaricus, Anabas testudineus, Channa* sp. *Oreochromis mossambicus* and *Mastacembeles armatus.* On the other hand, several marine forms were recorded from the backwaters during the summer season. *Rhinobatus halavi, Congresox talabonoides, Lobotes surinamensis, Acanthurus strigosus, Eleotris fusca, Lepturocanthus savala, Platax orbicularis* etc., were also recorded from certain systems, though in stray numbers. Thus it is clear that in the backwaters majority of the fish fauna have been recruited from the adjoining Arabian Sea, which shows that the fishery is mainly, depend on the ingress from the sea.

Of the 94 species recorded 67 are commercially important (Jhingran, 1975) and 12 contribute substantially to the fisheries (Table 34.3). The pearl spot, mullets, Lutjanids among fishes and *Penaeus indicus* are the most predominant forms having considerable market value and contribute appreciably to the landings. None of the commercially important groups belong to the truly estuarine. The fishery is mainly seasonal as it depends on the recruitment from the sea. *Etroplus* spp. *Nandus* sp. is available for all the seasons. *Valamugil seheli* and *V. speigleri* are available round the year. Among the Mullet, *Valamugil seheli* is the most common form. The major groups/species constituting the fishery with their relative contribution were estimated. *Etroplus suratensis* (2.0-13.9%), *Penaeus indicus* (2.5-29.6%) and *Metapenaeus monoceros* (1.0-8.2%) represented in the total catch. The mullets (1.5-16.5%), Lutjanids (1.1-11.4%), carangids (1.0-5.2%), the tiger perch *(Therapon jarbua),* the reef cod *Epinephalus* spp., *Sphyraena jello* etc. the *Gerres filamentosus* (1.0-6.1%) and the flathead *Platycephalus indicus* also formed a commercially important species in the fishery. The less known commercial fishes like silver bellies *(Leiognathus* spp.), the half beaks, the marine catfish *(Tachysurus* spp.), the anchovies *(Stolephorus* sp.), the flat fishes and the Ambassids gained importance due to their bulk contribution to the fishery.

Among the prawns, the most significant contributor to the fishery was *Metapenaeus dobsonii,* to the tune of 10.0-53.4 % to the total landings. *Penaeus indicus* also considerably contributed to the fishery (2.5-29.6 %) due to its attractive prize followed by *M. monoceros* (1.0-8.2%).

The crab fishery also substantially contributed to fishery (appro. 1100 t; 2.6-11.1 %) by the spotted crab *(Portunus sanguinolentus),* the blue crab (*P. pelagicus)* and the mud crab *(Scylla serrata)* serving both the domestic and international markets.

The total landings from different backwaters varied from 96.8 t (Mahe) to 2899 t (Ashtamudi; Table 34.3). The average yield per ha varied from 410 kg at Anchuthengu to 2747.3 t from Azhicode estuary. The low yield per unit area at Ashtamudi was due to its large area (6424 ha). The high density fish/prawns population at Azhikode is due to its wide access to the sea and the narrow stretch of the backwater spread. The system was well filtered by

Table 34.3.

Finfish and shellfish recorded from selected backwaters during 1996-97

Sl. No.	Species Name	Habitat	Backwaters									
			KDK	ANJ	AST	KYM	AZK	CTV	PNI	MHE	VPM	NEL
1	2	3	4	5	6	7	8	9	10	11	12	13
1.	*Rhinobats halavi (Day)*	M						*				
2.	*Megalops cyprinoides (Broussonet)*	E	**	**	**	**	**	**	**	**	**	**
3.	*Anguilla bengalensis (Gray)*	R	*	*	*	*		*	*	*	*	*
4	*Ophichthys altipinnis (Kaup)*	E										
5	*Congresoz talabonoides (Bleeker)*	M			**	**		***		**		
6	*Nematalosa nasus (Bloch)*	M	**	**	**	**	**	**	**	**	**	**
7	*Pellona ditchela (Valenciennes)*	M	**	**	**	**	**	**	**	**	**	**
8	*Dussumieria hasseeltii (Bleeker)*	M			*			*	*			
9	*Stolephorus indicus (van Hasselt)*	M	***	***	***	***	***	***	***	***	***	***
10	*Thryssa mystaz (Schneider)*	M	***	***	***	***	***	***	***	***	***	***
11	*Chanos chanos (Forskal)*	E	**	**	**	**	**	**	**	**	**	**
12	*Puntius filamentosus (Valenciennes)*	R	***	***	***	***	***	***	***	***	***	***
13	*Puntius sarana (Hamilton-Buchanan)*	R	**	**	**	**	**	**	**	**	**	**
14	*Labeo dussumieri (Valenciennes)*	R				*	**		**			
15	*Mystus malabaricus (Jerdon)*	R				**	*		**			
16	*Horabagrus brachysoma (Gunther)*	E	*			*		*		*	*	
17	*Chaca chaca (Hamilton-Buchanan)*	R										
18	*Anus platystomus (Day)*	E	****	****	****	*****	*****	*****	*****	*****	*****	****
19	*Arius rostratus (Day)*	E	***	***	***	***	***	***	***	***	***	***
20	*Hyporhamphus limbatus (Valenciennes)*	M,E	**	**	**	**	**	**	**	**	**	**
21	*Hemirhamphus canton (Day)*	M,E	**	**	**	**	**	**	**	**	**	**

1	2	3	4	5	6	7	8	9	10	11	12	13
22	*Hemirhamphus far (Day)*	M,E	**	**	**	**	**	**	**	**	**	**
23	*Strongylura strongylura (van Hasselt)*	M,E	**	**	**	**	**	**	**	**	**	**
24	*Tylosuruscrocodilus (Le Sueur)*	M,E	*	*	*	*	*	*	*	*	*	*
25	*Platycephalus indicus (Linnaeus)*	E,M	****	****	****	****	****	****	****	****	****	****
26	*Lates calcarifer (Bloch)*	E,M	*	*	*	*	*	*	*	*	*	*
27	*Ambassis commersoni (Cuvier)*	E,R	****	****	****	****	****	****	****	****	****	****
28	*Parambassis spp.*	R,E	****	****	****	****	****	****	****	****	****	****
29	*Epinephelus tauvina (Forsskal)*	M,E	**	**	**	**	**	**	**		**	**
30	*Epinephelus malabaricus (Schneider)*	M,E	**	**	**	**	**	**	**		**	**
31	*Terapon jarbua (Forsskal)*	M,E	**	**	**	**	**	**	**	**	**	**
32	*Sillago sihama (forsskal)*	M,E	****	****	****	****	****	****	****	****	****	****
33	*Caranx sexfasciatus (Quoy & Gaimard)*	M	***	***	***	***	***	***	***	***	***	***
34	*Caranx nigripinnis (Bleeker)*	M,E	**	**	**	**	**	**	**	**	**	**
35	*Leiognathus equulus (Forsskal)*	M,E,R	****	****	****	****	****	****	****	****	****	****
36	*Leiognathus decorus (Le Vis)*	M	****	****	****	****	****	****	****	****	****	****
37	*Leiognathus splendens (Cuvier)*	M	****	****	****	****	****	****	****	****	****	****
38	*Lutjanus argentimaculatus (Forsskal)*	M,E	**	**	**	**	**	**	**	**	**	**
39	*Lutjanus johni (Bloch)*	M	**	**	**	**	**	**	**	**	**	**
40	*Lutjanus fulviflamma (Forsskal)*	M,E	**	**	**	**	**	**	**	**	**	**
41	Lobotes Surinamensis *(Block)*	E,M							*			*
42	*Gerres filamentosus (Cuvier)*	E,M	****	****	****	****	****	****	****	****	****	****
43	*Gerres oyena (Forsskal)*	M,E	**	**	**	**	**	**	**	**	**	**
44	*Monodactylus argenteus (Linnaeus)*	E										
45	*Drepane punctatus (Linnaeus)*	M			**	**						
46	*Scatophagus argus (Linnaeus)*	E,R	***	***	***	***	***	***	***	***	***	***

1	2	3	4	5	6	7	8	9	10	11	12	13
47	*Acanthurus bleekeri (Gunther)*	M,E	**	**	**	**	**	**	**	**	**	**
48	*Acanthurus strigosus (Bennett)*	M	..	..	..	..	..	..	..	..	..	..
49	*Nandus nandus (Hamilton-Buchanan)*	R	....	....	....	....	....	....	....	....	....	....
50	*Nandus marmoratus (Hamilton-Buchanan)*	R,E			*	*						
51	*Etroplus suratensis (Bloch)*	E,R	....	....	....	....	....	....	....	....	....	....
52	*Etroplus maculatus (Bloch)*	E,R	....	....	....	....	....	....	....	....	....	....
53	*Oreochromic mossambicus (Peters)*	R	....	....	....	....	....	....	....	....	....	....
54	*Liza tade (Forsskal)*	M,E	....	....	....	....	....	....	....	....	....	....
55	*Liza parsia (Hamilton-Buchanan)*	M,E	....	....	....	....	....	....	....	....	....	....
56	*Migil cephalus (Linnaeus)*	M,E,R	**	**	**	**	**	**	**	**	**	**
57	*Valamugil seheli (Forsskal)*	M,E	....	....	....	....	....	....	....	....	....	....
58	*Valamugil speigleri (Bleeker)*	M,E,R	***	***	***	***	***	***	***	***	***	***
59	*Sphyreena jello (Cuvier)*	M	***	***	***	***	***	***	***	***	***	***
60	*Gossogobius giuris (Hamilton-Buchanan)*	M,E,R	***	***	***	***	***	***	***	***	***	***
61	Oxyurichthus Microlepsis (Bleeker)	M,E	***	***	***	***	***	***	***	***	***	***
62	*Oxyurichthys tentacularis (Valenciennes)*	M	***	***	***	***	***	***	***	***	***	***
63	*Oxyurichthys formosanus (Nichols)*	M,E			**	**	*					
64	*Eleotris fusca (Day)*	E,R	*						**		*	
65	*Lepturacanthus savala (Cuvier)*	M,E			*	*			*		*	*
66	*Anabas testudineus(Bloch)*	R	*			**	*		**			
67	*Channa marulius (Hamilton-Buchanan)*	R	*		*	**			**			
68	*Channa sp.*	R	*			**			**			
69	*Macrognathus guentheri (Day)*	R	*		*	*			*			

1	2	3	4	5	6	7	8	9	10	11	12	13
70	*Mastocembelus armatus (Lacepede)*	R	*	*	*	*	*	*	**	*	*	*
71	*Pseudorhombus javanicus (Bleeker)*	E,M	**	**	**	**	**	**	**	**	**	**
72	*Pseudorhombus arsius (Hamilton-Buchanan)*	M,E	*	*	*	**	*	**	*	*	*	*
73	*Cynoglossus cynoglossus (Ham-Buch)*	M,E	**	**	**	**	**	**	**	**	**	**
74	*Cynoglossus punticeps (Richardson)*	M	**	**	**	**	**	**	**	**	**	**
75	*Cynoglossus lingua (Hamilton-Buchanan)*	M,E	**	**	**	**	**	**	**	**	**	**
76	Euryglossa orientalis (Bloch & Schneider)	E	***	***	***	***	***	***	***	***	***	***
77	*Triacanthus biaculeatus (Bloch)*	M			**	*			*		*	*
78	*Tetraodon leopardus (Day)*	M,E	*	*	*	*	*	*	*	*	*	*
79	*Teuthis vermiculata (Day)*	M,E	**	**	**	**	**	**	**	**	**	**
80	*Teuthis java* (Day)	M	**	**	**	**	**	**	**	**	**	**
81	*Stromateus cinereus*	M	**	**	***	**	**	**	**	**	**	***
82	*Platax orbicularis (Forsskal)*	M										*
Prawns												
83	*Macrobrachium rosenbergii*	R,E		**	**	**	**	*	*		*	*
84	*Macrobrachium idella*	R,E	**	**	**	**	**	**	**	*	**	**
85	*Macrobrachium lamerrei*	R,E	**	**	**	**	**	**	**	*	*	**
86	*Penaeus indicus*	M,E										
87	*Penaeusmonodon*	M	***	***	***	***	***	***	***	***	***	***
88	*Penaeus semisulcatus*	M	**	**	**	**	**	**	**	*	*	*
89	*Metapenaeus monoceros*	M,E	****	****	****	****	****	****	****	****	****	****
90	*Metapenaeus dobsonii*	M,E	****	****	****	****	****	****	****	****	****	****
Stomatopod												
91	*Oratosquillanepa*	M,E	****	****	****	****	****	****	****	****	****	****
Crabs												
92	*Scylla serrata*	M,E	****	****	****	****	****	****	****	****	****	****
93	*Portunus pelagicus*	M,E	****	****	****	****	****	****	****	****	****	****
94	*Portunus sanguinolentus*	M,E	****	****	****	****	****	****	****	****	****	****

**** Abundant ***Moderate **Rare *Occasional

the Stake net (445 units) that brought huge quantity of *M. dobsonii, M. monoceros, P. indicus* and a whole lot of less commercial species. The Anchuthengu backwater brought less catch due to the denial of direct access to the sea and being a highly strained ecosystem resultant to pollution from coconut husk retting. The average value for the whole system was 630.1 kg ha^{-1} yr^{-1}.

The nets used in different backwaters are Gill net, Cast net, Seine net, Stake net, Chinese dip net, Scoop net, Ring net, Trawl net, Trap net and Hook and lines. Of these, cast nets and seine nets are commonly used; the latter is mainly used for prawn fishing. Occasionally anchovies, chandids and gobiids are also getting. Dragnets also used in the shallow region, which destroys large number of young ones of *Etroplus suratensis,* the most commercially important species in the state. Stationary dip nets installed on wooden frames popularly known as Chinese dip net are also used. The different gear observed for the fishing operation during the survey was broadly categorised into 10 different types. The gear enumeration indicates that the density (No. km^{-2}) different systems ranged from 52 numbers in Ashtamudi to 174 in Mahe. On an average, one gear unit has different composition and days of operation roughly averaging for the whole year. Operation of stake nets and Chinese dip nets are influenced directly by the tidal strength. Hence their operation is generally restricted to a few days adjacent to full moon and new moon.

The CPUE exhibited wide variation from gear to gear (Table 34.4), but a general trend for most the backwaters. The high rate CPUE for the Stake net is largely offset by the limited days of operation and the low market value fish forming bulk of the catch. Seine net too brought high CPUE, but again the catch composition is composed largely of small fishes, young fishes and low priced *M. dobsonii.* The catch from this gear was subjected to wide variation.

Table 34.4. Average CPUE and percentage contribution by different gear to the total landings in the backwaters during 1996-96.

Gear	CPUE (kg $unit^{-1}$ day^{-1})	Contribution to total landings (%)
Gill net	3.9-12.1	26.31
Cast net	2.5-6.8	10.94
Seine net	11.3-744.7	17.76
Stake net	7.5-18.1	29.39
Chinese dip net	5.6-10.2	11.6
Hooks and line	2.3-10.3	2.13
Scoop net	2.3-2.9	0.89
Ring net	1.0-5.6	0.91
Trap net	3.8-8.7	0.46

The general observation is that the fishery of the backwater is less dependent on the local species and more dependent on the migratory marine species like the prawns, crabs, catfishes, silver bellies, baracudas, anchovies, carangids, perches and so on. However, considerable contribution has done by the local resident species like *Etroplus suratensis* and mullets etc.

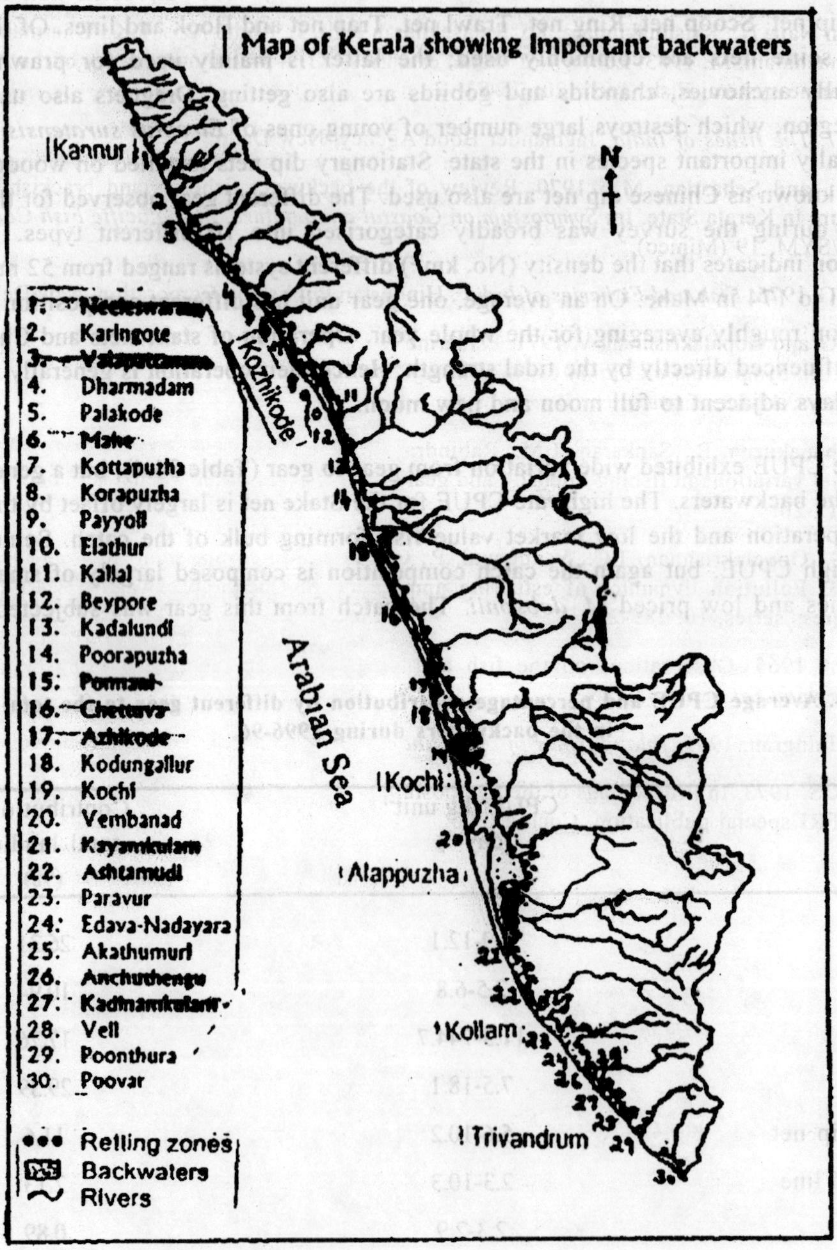

Fig. 34.1 Map of Kerala showing important backwaters

ACKNOWLEDGEMENTS

The authors are thankful to the Director, Central Inland Capture Fisheries Research Institute, Barrackpore for his constant encouragement for the preparation of the paper.

REFERENCES

Balakrishnan Nair, N., Krishnakumara, K., Rajasekharan Nair, J., Abdul Azis, P.K., Dharmaraj, K. and Arunachalam, M. 1983. Ecology of Indian estuaries- XI. A preliminary survey of the Ashtamudi estuarine system. Fish. Tech. 20: 75-83.

Day, F. 1967. *The fishes of India,* Jagmander Bood Agency, New Delhi.

George, A.I., and Sebastian, M.J. 1970. Review of the backwater fisheries and brackishwater fish culture in Kerala State. In: *Symposium on Coastal aquaculture, Indo-Pacific Fish Coun.* IPFC/ C70/SYM, 19 (Mimeo).

Jhingran, V.G. 1975. *Fish and Fisheries of India* , Hindustan Publishing Corporation, New Delhi, India.

Jhingran, V.G. and Gopalakrishnan, V, 1971. Estuarine fisheries resources of India in relation to adjacent seas. In: *Symposium on Indian Ocean and adjacent seas, Cochin.* Marine Biological association of India, 12-18, January.

Madhusoodhanakurup, B., Sankaran, T.M., Rabindranath, P. and Sebastian, M.J. 1993. Seasonal and spatial variations in fishing intensity and gear-wise landings of the Vembanad lake. Fish. Tech. 30: 15-20.

Nair, K.K.C., Gopalakrishnan, T.C., Venugopal, P., George, P.M., Jayalakshmi, K.V. and Rao, T.S.S. 1983. Pollution dynamics of estuarine amphipods in Cochin backwaters. Marine ecology progress series, 10: 289-295.

Shetty, H.P.C.1965. Observations on the fish and fisheries of the Vembanad backwaters, Kerala. *Proc.Acad.Sci.India,* 35:115.

Thalwar & Jhingran, 1991. *Inland fishes of India and adjacent countries* 1&2: 1-1158.

Thampi, P.R.S. 1973. In Proceedings of the Symposium on Living Resources of the Seas Around India, CMFRI special publication, Cochin, 390.

Haemato-Biochemical Studies On Some Economically Important North Indian Fishes. III. On The Sea Sonal Variation Of Organic Metabolite - Glucose

S. K. Singh, K.N. Srivastava* And Amar Kumar**

Department of Zoology, K.N.P.G. College Tejgaon, Raebareli, 229001, U.P.

**Fisheries Section, Department of Zoology, F.G. College, Raebareli- 229001, U.P.*

***The Pathological Laboratory, Kutchery Road Raebareli- 2290001 U.P.*

ABSTRACTS

Present paper holds, studies on the seasonal variation of blood glucose levels in *L. rohita, C, batrachm & C. striatus* under north Indian climatic conditions of a natural and uncontaminated habitat. During present work seasonal variation in blood glucose levels was frequently noted in all sample fishes. All the samples revealed highest values in the January (winter phase) owing to hyposecretion of insulin and hypersecretion of glucagon. Beyond January, elevated levels of blood glucose were also noted in May (summer phase) in all the samples owing to rapid glycogenolysis mechanism due to increased BMR at higher temperature and also due to prespawning phase of the sample fishes appearing in May. Spawning phase in July also shows higher levels of blood glucose. In major carps the lowest values were recorded in September while in cat fishes and live fishes, the lowest values were noted in - November (post spawning phase) owing to exhaustion of liver glycogen.

Key words: ***Fresh water fishes, blood glucose, seasonal variation.***

INTRODUCTION

The state of any biotic component of environment can be expressed by the set of numerical values of its variables at that moment. The biochemical constituents of any living

system in a normal range play vital role in sustaining life and maintaining good health. Fishes, the leading creatures of aquatic ecosystem, show a number of anabolic and catabolic changes associated with the physiological processes, by their serum chemistry profiles (Wallace, 1985). Mc.carthy *et al.* (1973) prophesied that in future fish blood biochemistry will play vital role as diagnostic tools. But if more work is not performed in formulating normal ranges, this prophesy will be scant. Present series (Singh & Srivastava, 1999; 2002) of work is an humble contribution to the same from the north Indian scenerio as fishes bear poikilothermic nature and the Indian literature lacks the same in comparison to abroad (Hille, 1982; Sandness *et al.*, 1988). Present study embraces, first the monitoring of the objective parameter in the leading fishes groups in the diffrent seasons to evaluate the normal ranges and secondly the analysis of these findings in the light of previous scattered works (Folmar *et al.*, 1992; Saxena & Chauhan 1993 C; Rauthan & Grover, 1994). Undoubetedly such studies will strengthen the plateform of fresh water fish pathology, offer better application since widely used today in toxicological screenings (Anandhi *et al.*, 2002; Shankar & Kulkarni, 2002; Saxena & Sheth, 2002; Sherry & Abidi, 2002 : Talapatra *et al.*, 2002) and play vital role in restoration of depleating aquatic protein for mankind owing to anthropogenically stressed habitat.

MATERIALS & METHODS

For present study three fishes were selected related to three chief fresh water fish groups (Pandy, 1983); *Labeo rohita* (common name- Rohu) from major carps; *Clarias batrachus* (common name - mangur) from cat fishes and *Channa striatus* (common name - saur) from live fishes. Fish pond, fish culture and capture methods, anaesthetisation pattern, mode of blood withdrawal, sampling months, mean length and weight,estimation of objective parameter in sample fishes were same as mentioned by Singh (1997). For estimation of blood sugar Sodium flouride was used as anticoagulant to cease glycolysis. Estimation were conducted at 37°C as suggested by Casillas *et al.* (1982) fit for fish blood. Blood glucose was estimated randomly by glucose oxidase method, kit- span diagnostics, code no. - 25940.

RESULTS & DISCUSSION

During present work seasonal variation in glucose level was frequently observed in all sample fishes. In *L. rohita* peak values were observed in January and then in May while lowest values were noted in September. In C *batrachus and C. striatus* peak values were noted in January followed by in July then May while lowest values were observed in the month of November (Fig-35.1, Table-35.1)

Table-35.1
Haemato biochemical values noted in sample fishes (mg/100 ml of blood)
***All values are mean ± S.e. of 3 observations**

Month	L.rohita	C. batrachus	C. Striatus.
November	54.61±2.33	72.16±5.29	77.54±4.18
January	86.38±4.57	98.28±5.20	112.54±10.62
March	60.46±3.25	70.44±4.22	84.00±4.38
May	80.64±1.38	92.23±4.64	100.00±9.12
July	62.20±3.14	94.32±4.26	106.40±6.35
September	48.53±2.46	81.30±3.22	93.70±.435

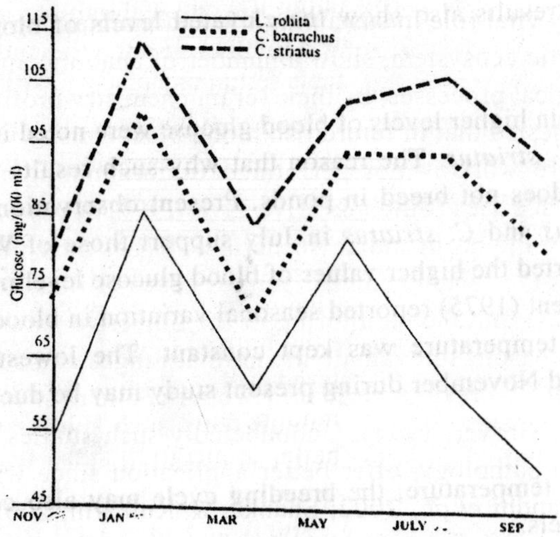

Fig. 35.1

Graph showing seasonal variation of mean blood glucose levels in *L.rohita*, *C. batrachus* and *C. striatus*.

During present work of seasonal variation in blood glucose levels, significant inter-relationships were observed in between water temperature, breeding cycle and glucose levels. In all the examined fishes the blood glucose levels were at peak in winter (January). This finding supports the observations of Pavlovic (1968), Umimnger (1969 b, c, 71 c, 72 a),Umminger and Bair (1972), Joshi (1973), Block (1974), Leach and Taylor (1977), Lane (1979), Folmar et al. (1992), Saxena & Chauhan (1993c), who reported the same, but did not support those of Haider (1970 a), Rauthan & Grover (1994) who reported lower values of blood glucose levels in winter times.

The purpose of rising blood glucose at lower temperature is to permit survival in cooler conditions. This may be due to the less utilization of glucose at lower temperature due to slowing down of whole metabolic-processes/ BMR. Freed (1965) and Houstan (1973) mentioned that depressive effect of lower temperature on fish metabolism are partially compensated by the increase of not only certain enzyme activities but there is increase in certain hormonal secretions also as thyroxine, which is responsible for maintaining the tempo of life, attains a maximum level in winter (Osborn *et al.*, 1978). Umminger (1971 C) who made an extensive work on blood glucose level in fishes, noted that hyperglycemia at lower temperature is the effect brought about by a decrease in insulin productions. Umminger and Bair (1973) further revealed that at lower temperature the beta cells of pancreas, which produce insulin, do indeed atrophy while the alpha cells, which produce glucagon become hypertrophy. Thus it is clear that maximum values of blood glucose levels in winter times in fishes are due to hyposecretion of insulin and hypersecretion of glucagon.

After lowering in March blood glucose again elevates in summer (in May) in all sample fishes. This finding supports the observations of Wedemeyer (1973), Grigo (1975), Rauthan & Grover (1994) who noted higher values of blood glucose at higher temperatures. This may be due to the fact that increasing water temperature enhances the metabolic activities, hence increased glycolysis compensated by rapid glycogenolysis mechanism appears at higher temperatures. Rapid glycogenolysis elevates glucose level in blood. Consistent with findings of Folmar *et al.* (1992), our results also show the elevated levels of blood glucose in pre-spawning phase in May.

However after May again higher levels of blood glucose were noted in July as spawning period in *C. batrachus and C. striatus.* The reason that why such results don't appear in *L. rohita,* is perphaps that this does not breed in ponds. Present observations of higher blood glucose levels in *C. batrachus* and *C. striatus* in July support those of Wendt and Ericson (1972), Joshi (1973) who reported the higher values of blood glucose level in spawning period. Pavlovic (1968); Murat & Parent (1975) reported seasonal variation in blood glucose levels in fishes even when the water temperature was kept constant. The lowest values of blood glucose level in September and November during present study may be due to the exhaustion of liver glycogen.

Thus in fishes, besides temperature, the breeding cycle may also cause the seasonal variation in blood glucose levels.

Table 35.2

Levels of blood glucose (in mg/ 100ml of blood) in some Indian fresh water fishes.

Sn.	Fish Species	Range	Mean	Source
1.	*Anabas testiedineus*	—	125.25	Akela *et al.* (1995)
2.	*Barilius bendelisis*	49.0 - 65.0	56.86	Raunthan & Grover (1994)
3.	*Channa punctatus*	78.00-91.60	84.36	Joshi (1973)
4.	*Channa punctatus*	34.00-40.00	—	Agarwal (1992)
5.	*Channa punctatus*	—	69.26	Akela *et al* (1995)
6.	*Channa striatus*	—	66.14	Akela *et al.* (1995)
7.	*Cyprinus carpio*	—	57.00	Saxena & Chauhan (1993b)
8.	*Clarias batrachus*	60.90-72.80	65.04	Joshi (1973)
9.	*Clarias batrachus*	—	90.14	Akela *et al.* (1995)
10.	*Heteropneustes fossiles*	—	92.25	Akela *et al.* (1995)
11.	*Heteropneustes fossiles*	—	57.00	Singh & Reddy (1990)
12.	*Hetero Pneustes fossiles*	—	57.00	Saxena & Chauhan (1992c)
13.	*Labes rohita*	—	86.80	Dalela *et al.* (1980)
14.	*Lobes rohita*	42.00-58.80	48.86	Joshi (1973)
15.	*Mystus vittatus*	—	84.60	Bansal *et. al.* (1979)
16.	*Rita rita*	54.18-86.37	67.09	Pandey (1976)
*	*Channa striatus*	77.54-112.54	95.69	Present Work
*	*Charias batrachus*	72.16-98.28	84.78	Present Work
*	*Labeo rohita*	48.53-86.38	65.47	Present Work

— Data not available.

ACKNOWLEDGEMENTS

The first author is thankful to the Manager & Principal,K.N. Post graduate college Tejgaon, Raebareli, U.P. for materialising the study through financial assistance since they have provided an opportunity to serve the institution.

REFERENCES

Agarwal, S.K. 1992. Sublethal effect of Mercuric Chloride on some biochemical parameters of the blood in an air breathing fish *Channa punctatus* (Bloch). J. Env. Biol. 13 (2): 127-133

Akela, B.P., Kumari, I .& Ritu, K. 1995. Effect of bacterial ulcer on biochemical constituents ofcertain fishes. Env. & Eco. 13 (2) : 378 - 380.

Anandhi, U., Laxmi D., Venkataramnana, N & Murthy, P.S. 2002 Toxic effect of Malathion on the liver, ovary and kidney of fresh water Gobied fish *Glossogobius giuris*- biochemical study. Proceed. 89th sesson of Ind. Sci. cong. Lucknow University, Zoology section, Abs. No. 142.

Block, R.M. 1974. Effect of acute cold shock on the channel cat fish. AEC symp. series. 32: 109-118.

Bansal, S.K., Verma, S.R., Gupta, A.K. & Dalela, R.C. 1979 b. Physiological disfunction of the Haemopoietic system in a fresh water teleost *Labeo rohita*, following chronic chlordane exposure, Part II. Alterations in certain organic components and serum electrolytes, Bull.Environ. Contam. Toxicol. 22 : 674 - 680.

Casillas, E., Sundquist, J. & Ames, W.B. 1982. Optimization of assay conditions for and the selected tissue distribution of alanineamino transferase and aspartate aminotransferase of English Sole, *Parophrys ventulus girard.* J. fish. Biol. 21 : 197 - 204,'

Dalela, R.C., Rani, S.,Kumar, V. & Verma, S.R. 1980. In Vivo haematological alterations in a fresh water teleost *Mystus vittatus* following subacute exposure to pesticidies and their combinations. J. Env. Biol. 2(2) : 79 - 86.

Freed, J.M. 1965. Changes in activity of cyto chrome oxidase during the adaptation of gold fish to different temperatures. Comp.Biochem. Physiol. 14 : 651-659.

Folmar, L.C., Moody, T., Bonomilli, S. & Gibson, J. 1992. Annual cycle of blood chemistry param etersin striped mullet *(Mugilcephalus. L.)* and pin fish *(Lagodon rhomboides L.)* from the Gulf of Mexico. J. Fish. Biol. 41 : 999-1011.

Grigo, F. 1975. How much is carp *(Cyprinus carpio)* streesed by temperature? 1. blood composition with a special look at the serum electrolytes. Zool. Anz. Zena.8:215-233.

Haider, G. 1970 a. Hamatologische Beobachtungin an Regenbogen forellen *(Salmo gairdneri. Rich.)* . II. Derblutzuckerspiegel Z. Fischerei NF, 18 : 209-216.

Houstan, A.H. 1973. Environmental temperature and the body fluid system of the teleost. In Responses of fish to Environmental changes. Thomas spring field. Illinois 4 : 87-162.

Hille, S. 1982. A literature review of the blood chemistry of rain bow trout, *Salmo gairdneri Rich.* J. Fish. Biol. 20 : 535- 569.

Joshi, B.D. 1973. Studies on blood of fresh water fishes of India. Ph. D./thesis, Deptt. of Zoology, University of Lucknow.

Leach, G.J. and Taylor, M.H. 1997. Seasonal measurements of serum glucose and serum cortisol in a natural population of *Fundulus heteroclitus.* Comp. Biochem. Physiol. 56 A : 217 - 223.

Lane, H. C. 1979. Progressive changes in haematology and tissue water of sexually mature *trout, Salmo gairdneri* (Rich) during the autumn and winter. J. fish biol. 15 : 425 - 436.

Mc. Carthy, D.H., Stevenson, J.P. and Robberts, M.S. 1973 Some blood parameters of the rainbow *trout Salmo gairdneri (Rich)*. J. fish biol. 5:1-8.

Murat, J.C. and Parent J .P. 1975. Effect of temperature acclimation on growth and carbohydrate metabolism in carp. Preliminary data, Chaiers lab, Montereau 2 : 23 - 24.

Osborn, R.H., Simpson, T.H. and Youngson, H.F. 1978. Seasonal and diurnal rhythms of thyroidstatus in rainbow trouty, *S. gairdneri (Rich)*. J. Fish. Biol. 12:531-534.

Pavlovic, V. 1968. Seasonal dynamics of carbohydrate metabolism at two kinds of freshwaterfish. Z. vergl. Physiol. 59 : 72 - 77.

Pandey, A.K. 1976. Morphophysiological studies on the blood of some fresh water fishes of Uttar Pradesh. Ph.D.thesis, Deptt. of Zoology, University of Lucknow.

Pandey, K.C. 1983. Concept of Indian fisheries. Meera Publication Co. Faizabad U.P.

Rauthan, J.V.S. and Grover, S.P. 1994. Effect of some eco- physiological conditions on haemato logical values of Hill stream fish *Barilius bendelisis* (Ham.) of Doon Valley.U.P.Z. Zool 14(2): 147 - 150.

Sandnes, K., Lie, O. and Waagbo, R. 1988. Normal ranges of some blood Chemistry parameters in adult farmed Atlantic Salmon *Salmo solar.* J. fish. biol32: 129-136.

Singh, H.S. and Reddy, T. 1990. Effect of copper sulphate on Haematology Blood chemistry and Hepato somatic index of an Indian cat fish, *H. fossilis* (Bloch) and its recovery.Ecotoxicology and Environmental Safety 20 : 30 -35.

Saxena, K.K. and Chauhan R.R.S. 1993. Effect of Starvation on the blood of fish *Cyprinus Carpio* (Linn). Trends in life Scineces (India).8 (1) : 1-4.

Singh, S.K.1997.Taxonomy, haematology of fish fauna of Raebareli with special reference to metal and pesticide pollutants .Ph.D.thesis Kanpur University Kanpur.

Singh, S.K. and Srivastava K.N. 1999. Haemato- biochemical studies on some economically important north Indian fishes I.On the seasonal variation of TEC, Hb and PCV. Uttar Pradesh J. Zool 19 (1) : 53 - 57.

Singh S.K. and Srivastava K.N. 2002 .Haemato- biochemical studies on some economically important North Indian fishes II. On the seasonal variation of TLC, DLC & ESR. Paper presented at Ind. Sci. Cong. 89th Sesson, Zool Sect. University of Lucknow.

Saxena K.K. and R.R.S. Chauhan, 1993 C. Haematological responses to hypothermic variations in a carp *Cyprinus carpio* (L). Him. J. Env. Zool Vol7:194-197.

Saxena K.K. and Nishtha Sheth. 2002. Fenvalerate induced Haematological strees in a fresh water fish *Channa punctatm.* Proceed. 89th Sesson of Ind. Sci. Cong.Zoology Section, Abs. no 144- University of Lucknow.

Sherry, P.M. and Abidi. Z.Z. 2002. Erythrocytic and Leuko cytic responses to lead and cadmium poisoning in a fresh water teleost *Channa punctatus* (Bloch) Proceed 89th sesson of Ind Sci. Cong. Zoology section ABS No. 171. University of Lucknow.

Talapartra, S.N., Ranu Naskar, George, P. and Sen, N.S. 2002. Impact Of chromium thiodan and Ekaline FI toxicity on the Haemato biochemical parameters of *C. batrachus* (Bl.).Proceed. 89th sesson of Ind Sc. Cong. Zoology Section, Abs No. - 165, University of Lucknow.

Umminger, B.L. 1969b. Physiological studies on super cooled kill fish *(Fundulus heteroclitus)* II. Serum organic constituents and the problem of super cooling J. Exp. Zool. 172: 409 - 423.

Umminger, B.L. 1969c. Physiological studies on super cooled kill fish *(Fundulus heteroclitus)*. III. Carbohydrate metabolism and survival at sub zero temperatures. J. Exp Zool 173 : 159 - 174.

Umminger, B.L. 1971c. Lack of pituitary involvement in the cold - induced hyperglycemia of the kill fish, *Fundulus hiteroclitus*. J. Exp. Zool 27 : 701-702.

Umminger, B.L. 1972a. Physiological studies on super cooled kill fish *(Fundulus eteroclitus).I.V.* Carbohydrate metabolism in hypophysectomised kill fish at sub zero temperatures. J. Exp. Zool.181 : 217 - 222.

Umminger, B.L. and Bair, R.D. 1973. Role of islet tissue in the cold - induced hyperglycemia of the kill fish *Fundulus heteroclitus*. J. Exp. Zool. 183 : 65 -70.

Wendt,C. and Ericson 1972. Blood glucose in hatchery reared atlantic salmon *(Salmo salar L.)* following exercise in winter andsumrner. Rep.Inst. Fresh Wat. Res. Drottningholm 46:167-184.

Wedemeyer, G. 1973. Some physiological aspects of sublethal heat strees in the juvenile steelhooc trout *(Salmo gairdneri)* and coho salmon *(Oncorhynchus Kisutch)*.J.Fish.Res.Bd.Cam.30:831-834.

Wallace, R.A. 1985.Vitellogenesis and oocyte growth in non-mammalian vertebrates. In: Developmenta Biology (Ed. L. W. Browder) Plenum Press, Newyork.PP.127- 177.

CHAPTER 36

Impact of Dairy Effluents on the Behavioural Responses of Fish, *Glossogibus giuris* (Ham)

Krishna Ram H and M. Ramachandra Mohan
Department of Zoology, Jnana Bharathi, Bangalore University, Bangalore-56, (Karnataka) India.

ABSTRACT

The present experimental studies, revealed the effect of dairy effluents on the behavioural responses of fresh water edible fish, *Glossogibus giuris* (HAM). Today's world is more dependent on environment resources and development man's activities are continuously polluting all such resources. Lake water pollution is one such problem, which is gaining concern, especially in fast developing countries. Balaganahalli lake of Kolar (Karnataka) was highly polluted with dairy effluents, it's water are not suitable for either aquatic life nor agriculture. In this contest the behavioral responses of gobid fish has been taken and fish exhibited various behavioural responses like, locomotary, physiological, neurological, morbidity under the influence of dairy effluents. The present study also indicated the possible utility of the behavioural responses as rapid biomonitoring assessment technique for qualitative evaluation of various parameters of dairy effluents.

Keywords: *Dairy effluents, fish behaviour, Glossogibus giuris*

INTRODUCTION

The lake pollution is a serious water pollution problem, which needs concern. Fish behaviour has been an interesting area of study for aquatic biologists. The physical, chemical, biological analysis alone do not provide adequate information in respect of environment contamination. The altered environmental influence can be detected and evaluated by studying the behaviour of fish under appropriate condition. Study on behaviour of fishes under environmental stress was carried out by Furlinger and Beamish (1977), Wildish *et al,* (1977), Trivedy and Dubey (1978), Srinivas *et al.* (1984), Waterpaugh and Beitinger (1985), Nakarnura (1986), Vijayram *et al.,* (1991), Sen *et al.* (1991), Amudha *et al,* (1997) and B.K.Baruah & M.Das (2002). Review of literature indicated very limited information regarding the effect of dairy effluent on the behaviour of fish, therefore, the present experiment was carried out to find out behavioural responses of a freshwater fish exposed to dairy effluent.

MATERIALS AND METHODS

Mature healthy live fish *G. giuris* of size 11-13cm were procured locally and acclimatized in glass aquarium (size 30 x 45 x 30cm) for one month and fed with earthworms. The dairy effluent of Kolar (Karnataka) was collected from the point of discharge in plastic container and brought to laboratory. Eight fishes from control aquarium were transferred in to test aquarium containing dairy effluents the fishes were exposed to the effluent for a period of 2 hrs, under normal temperature and natural light and different data regarding behavioural responses were recorded.

RESULTS AND DISCUSSION

The result of the observations are given in Table 36.1. The experiment clearly revealed that the behaviour of fishes remarkably changed due to the influence of dairy effluent. The various locomotary responses exhibited by the fishes at initial stage of exposure include restlessness, erratic and faster swimming, abrupt change in position and direction, jumping and overall hyperactivity. The fishes showed surfacing tendency throughout the experimental period. The overall hyperactivity suggested their tendency to get out of the test aquarium containing dairy effluent. The dairy effluent perhaps caused irritation, inflammation and injuries to the skin, gills and buccal cavity tissues. The overall change in activities may be related to the alteration of metabolic pathway to yield energy under oxygen depleted environment. This observation were in conformity with earlier experiments in fish under different pollutant stress. Wildish *et al.* (1977), Ferlinger and Beamish (1977), Trivedy and Dubey (1978), Sen *et al.* (1991) and Baruah *et al.* (1998).

Table-36.1

Behavioural responses of fish *G. giuris*

Responses	1-l0min	10-20min	20-40min	40-80min	80-120min
1	2	3	4	5	6
(1) Locomotary					
I. Restlessness	Observed	Observed	Observed	Not Seen	Not Seen
II. Erratic Movement	Observed	Not Seen	Not Seen	Not Seen	Not Seen
III. Fast Movement	Observed	Not Seen	Not Seen	Not Seen	Not Seen
IV. Position change	Frequent	Observed	Observed	Observed	Not Seen
V. Direction Change	Observed	Observed	Observed	Not Seen	Not Seen
VI. Sudden Swimming	Occasionally	Occasionally	Not Seen	Not Seen	Not Seen

1	2	3	4	5	6
VI. Surfacing	Observed	Observed	Observed	Observed	Observed
VII. Jumping	Not Seen	Not Seen	Observed	Observed	Observed
VIII. Sustained Swimming	Not Seen	Not Seen	Not Seen	Not Seen	Not Seen
IX. Hyperactivities	Observed	Occasionally	Occasionally	Not Seen	Not Seen
(2) Neurological					
I. Jerks	Not Seen	Not Seen	Not Seen	Not Seen	Not Seen
II. Frieghtened	Not Seen	Not Seen	Not Seen	Observed	Observed
III. Balance Loss	Not Seen	Not Seen	Not Seen	Observed	Observed
(3) Physiological					
I. Gulping	Observed	Not Seen	Observed	Observed	Observed
II. Operacular Movement	Rapid	Rapid	Rapid	Rapid	Rapid
(4) Morbidity					
I. Reduced Swimming	Occasionally	Observed	Observed	Observed	Observed
II. Slow Movement	Observed	Observed	Observed	Not Seen	Observed
III. Stationary	Observed	Not Seen	Not Seen	Not Seen	Not Seen

The physiological responses of fish as evidenced by gulping and rapid opercular movement were indicative of respiratory inconveniences triggered by the absence of dissolved oxygen in the effluent. Similar findings in relation to respiration were also recorded by Panigrahi and Mishra (1978), Malyarevkya (1978), Watenpaugh and Beitinger (1985) and Sen *et al.,* (1991). The neurobehavioural changes relfected by loss of balance during the exposure period due to the influence on sensory-motor interaction of the nervous system. The effluent might act as neurotoxic leading to disorganization of nervous co-ordination. This findings were in the line of earlier report of Panigrahi and Mishra (1978) and Ram and Gopal (1991) in fish intoxicated by pollutants. In the later stages of experiment, the fishes showed morbidity responses like reduced activities may be explained by the infiltration of toxic component of the effluent into the body systems resulting inhibitory influences to a variety of bio-chemical mechanisms. Reduced activity of fishes under environmental stress were also recorded. Our results are concides with Panigrahi and Mishra (1978) and Malyarevskya (1978).

The present experiment clearly demonstrated behavioural changes in the *G. giuris* under dairy effluent stress. The different types of responses may be useful in obtaining qualitative data regarding the biotoxic nature of different industrial effluent. The behavioural responses in fish may be utilized as rapid effluent quality assessment technique in biomonitoring study by the dairy environmental workers to acquire immediate preliminary information in relation to quality of effluent or polluted water.

REFERENCES

Baruah, B.K; Das.M. 2002. Study on Behavioural Responses of Fish Heteropneustes fossilis exposed to paper mill effluent. Indian J. Environ. & Ecoplan. 6(2): 263-266.

Baruah, B.K; Baruah, D. and Das, M. 1996. Sources and characteristics of paper mill effluent. Environment and Ecology. 14:686-689.

Baruah, B.K.; Sengupta, S. and Das, M. 1998. Toxocological impact of paper mill effluent on muscle glycogen profile of *H. fossilis* (Bloch.) Gau. Univ.J.Sci.Golden Jubilee Volume. 1-5.

Farlinger, S. and Beamish, F.W.H. 1977. Effect 0f time and velocity increments of the critical speed of large mouth brass *(Micropterous salmoid)*. Trans. Ass.Fish. Soc. 106:436-439.

Malyarevskya. A.K. 1978. Specific and non specific changes induced in fish by various toxic agents. Hydrobiological Journal. 14:52-59.

Nakamura, F.1986. Avoidance behaviour and swimming activity of fish to detect pH changes. Bull. Environ. Contain. Toxicol. 37 L 808-SIS.

Panigrahi, A.K. and Mishra, B.N. 1978. Toxicological effect of mercury on a freshwater fish *Anabas scandens* (Cuv and Val) and their ecological implications. Environ. Pollut. 16: 31-39.

Ram, M.D. and Gopal, K. 1991. Neurobehavioural changes in freshwater fish *Channa punctatus* exposed to fenitrothion. Bull. Environ. Contain. Toxicol. 47: 455-458.

Sen, G; Behera, M.K. and Patel, P.N. 1991. Toxicity of Zinc to the fish *Channa punctatus* (Bloch) with behavioural, morphological and skeletal abnormalities. Environment and Ecology. 9: 1023-1027.

Srinivas, P.; Benerjee, G. and Rao, P. 1984. Preliminary observations on the effect of industrial effluents to certain animals. Indian J. Environ. Hlth. 26: 57-60.

INDEX